The Paramedic at Work

A Sociology of a New Profession

Leo McCann

OXFORD
UNIVERSITY PRESS

Great Clarendon Street, Oxford, OX2 6DP,
United Kingdom

Oxford University Press is a department of the University of Oxford.
It furthers the University's objective of excellence in research, scholarship,
and education by publishing worldwide. Oxford is a registered trade mark of
Oxford University Press in the UK and in certain other countries

©Leo McCann 2022

The moral rights of the author have been asserted

Impression: 1

Published in the United States of America by Oxford University Press
198 Madison Avenue, New York, NY 10016, United States of America

British Library Cataloguing in Publication Data

Data available

Library of Congress Control Number: 2021952511

ISBN 978-0-19-881636-2

DOI: 10.1093/oso/9780198816362.001.0001

Printed and bound by
CPI Group (UK) Ltd, Croydon, CR0 4YY

Preface

There had been warnings about this for decades. Risk assessments and preparedness audits buried deep in forgotten filestores. Principles, procedures, and colour-coded flowcharts describing the likely trajectories of a deadly new disease pandemic. A government simulation in October 2016, known as Exercise Cygnus, roleplayed the response of healthcare, local government, police, fire and rescue, and military organizations to a flu-like disease outbreak. It found major shortages and deficiencies in the pandemic response, with the health service overwhelmed. It predicted 400,000 deaths. The findings of Cygnus were deemed 'too terrifying' for public consumption and the report was initially supressed.[1]

Just over three years later a dangerous new virus broke out for real. The SARS-CoV-2 virus hit the UK extremely hard, across three giant waves of infection in the spring of 2020, the autumn/winter of 2020–2021, and again in the autumn/winter of 2021-2022. The National Health Service came close to being overwhelmed in the first two peaks. At the time of writing there have been more than fifteen million infections and over 15,000 deaths attributed to the Covid-19 disease.

Working conditions for almost all people were upended by regional and national lockdowns, travel bans, and social distancing measures; by bankruptcy and closure of facilities; and by job cuts and staff sickness absence. Healthcare was one of the most traumatic, exhausting, and risky work environments imaginable. Frontline healthcare professionals of all kinds, as well as support, logistical, and administrative staff, had little or no option of working from home. Instead, they continued working at the heart of a dreadful and prolonged crisis. News media tended to focus most of their attention on hospitals and on the doctors, nurses, and other clinicians treating critically ill patients in specialist Covid-19 isolation wards. The hospital became a scene of desperate bleakness. Staff were shattered by the workload, wary of the extreme risk of contracting the virus themselves,

---

[1] 'Exercise Cygnus uncovered: the pandemic warnings buried by the government', *The Telegraph*, 28 March 2019.

and emotionally drained by constant trauma coupled with the unbearable loneliness imposed by isolation measures. Patients were suffering and dying without the comfort of loved ones at the bedside.

A modern healthcare system is broad and complex, encompassing many other professions and sites. The clinical workers who are the focus of this book work out in the community, operating in a 'pre-hospital' working environment that is at once complex, dynamic, uncertain, and poorly understood. Clearly the first UK-wide lockdown of March–July 2020 was extremely challenging for frontline paramedics and for ambulance services, especially the crisis that enveloped care homes. But, as the pandemic evolved, some elements of working life for ambulance crews were not quite as dreadful as first feared. Paramedics and ambulance services felt the impacts of the Covid-19 pandemic in ways that were distinct and idiosyncratic. The general population was extremely cautious, heeding the government message to stay indoors and massively restricting interpersonal contact. Members of the public were especially reluctant to go to hospital, and thought long and hard before calling 999 and running the risk of being taken there by an ambulance crew. The broader NHS had also postponed or cancelled all non-urgent consultations and procedures, turning most of their attention to Covid patients. Everyday activities that often result in people getting injured or sick happened much less often. There were far fewer road traffic accidents. There were no entertainment or sports events for which emergency services would provide backup. There were no callouts to drunks or to people with injuries from fighting. Even the psychiatric calls dropped off in volume; many of those often driven by anxiety or depression to call 111 or 999 found ways to suppress that tendency.

Ambulances were mostly being dispatched, therefore, to Covid-related calls that were 'genuine' emergencies. Emergency funding provided for extra crews on shift, and some locations had support from the Fire and Rescue Service and the Army. Ambulance stations were busy with staff waiting for calls, meaning that—unusually—there was time for staff interaction, to share advice, and to 'decompress' after stressful and exhausting shifts. Each individual crew didn't receive as many calls as they usually would, but when a call did come in, it was typically serious and challenging. Patients treated by paramedics tended to be very sick indeed. There were thousands of calls to patients stricken with life-threatening Covid-related conditions such as pneumonia, hypoxia, and heart attacks. For a short time, the changed rhythm and activity during the pandemic reminded crews of 'the old days'; 12-hour shifts featuring only around four or five callouts, but each to a patient facing a genuinely dangerous health emergency.[2]

The first lockdown was eventually successful in suppressing infection rates, hospitalizations, and deaths, and UK society started to reopen in July 2020. But, predictably enough, the rates of Covid infection slowly crept back up throughout late summer and accelerated in autumn. With winter on the horizon, the healthcare economy was building towards disaster. By January 2021, the country was back into a lockdown almost as severe as the first one in March 2020. Covid-19 infection rates exploded, peaking at around 60,000 known infections per day. Hospitals were once again overwhelmed with coronavirus patients. Over a thousand Covid patients were dying every day. Calls to ambulance services now hit record levels, such as 8,500 calls per day to London Ambulance Service, instead of the usual 5,000–6,000.[3] All around the country, ambulances were stuck in long queues outside hospitals, unable to offload Covid patients for hours because of a lack of available beds inside. Hospitals had tried to keep routine and elective appointments going, only dropping this workload to accommodate the second wave when they had no other choice. The 'normal' calls to the ambulance service didn't drop off. The more recent lockdown was less strictly adhered to than the first. The population was also stressed, isolated, and exhausted by nine months of life in a pandemic, so the mental health calls came back. Virus outbreaks occurred within ambulance control centres and HQs, leading to prolonged illness, staff shortages, and work overload. This time around, none of the paramedics out on the streets were making allusions to 'the good old days'. Instead, they were, in the words of paramedic and author Jake Jones, 'frustrated, drained, [and] powerless to help'.[4]

Concerns started to circulate about the probable inadequacy of the personal protective equipment (PPE) issued to ambulance crews. It consisted of Level 2 PPE to be worn to all calls (surgical facemask, disposable gloves, and apron), and Level 3 PPE (FFP3 mask or respirator hood, fluid repellent overalls) to be used when crews carry out 'aerosol generating procedures' such as chest compressions. One paramedic described Level 2 to

---

[2] An official overview of the pandemic from the perspective of ambulance service employers is AACE (2021). For a personal blog from a paramedic, see 'On the front line', *UK Climbing*, 14 April 2020, available at: https://www.ukclimbing.com/articles/features/on_the_front_line-12718

[3] 'London Ambulance Service: 'We take thousands of calls a day – it's tough', *BBC News*, 18 January 2021.

[4] 'Even as the Covid crisis accelerates, paramedics like me see people taking risks', *The Guardian* 6 January 2021.

me as little more than 'a plastic pinny you'd use for making sandwiches'. Paramedics were seated with Covid-infected patients in the confined space of an ambulance interior, waiting for hours at a time as hospitals struggled to find space to admit new patients. Staff were warned not to break protocol and upgrade themselves to Level 3 in such situations. Doing so in any case would often seem inappropriate and would offend a paramedic's understanding of appropriate patient treatment—respirators or hoods make interaction extremely difficult, with one paramedic commenting on Twitter that the Level 3 gear in use in his Trust makes the wearer look like Darth Vader.[5] Many broke the rules and donned the gear. Some resorted to buying their own FFP3 masks from Screwfix.

By October 2020, around 1 in 12 paramedics had tested positive for Covid-19 and by January 2021, at least 18 frontline ambulance workers in the UK had died of the disease.[6] Mercifully, ambulance staff were finally inoculated by the end of January 2021, with vaccines providing a way out of the nightmare that they, and other NHS staff, had suffered for so long. The vaccines did not fully prevent new infections, but were very effective in breaking the link between infection and serious illness. A further, prolonged wave of infection in the autumn and winter of 2021 was less severe than the previous two. The virus became endemic, and a largely vaccinated society learnt to adapt to live with it without lockdowns and social distancing measures. Still, the NHS as a whole faced a terrible backlog of postponed patient care, and found itself facing drastic operational difficulties throughout the year even before it headed into its always-difficult winter months.[7]

I wrote about half of this book during the years 2020–21. I had planned to finish it long before 2020, and the research data upon which it is based predate the Covid-19 pandemic by several years. But I had to reflect on these contemporary events somewhere in the text. Even without the pandemic, a central theme of the book was of professionals facing stress and burnout, and of trying to serve the public in extremely trying circumstances and with limited support. Already struggling with

---

[5] Tweet by @ClipboardUK, 3 January 2021, available at:
https://twitter.com/ClipboardUK/status/1345893679244963841?s=20
[6] Tracey Nicholls, Chief Executive of the UK College of Paramedics, to Rt Hon Matt Hancock, Secretary of State for Health and Social Care, January 6, 2021. Available at: https://www.collegeofparamedics.co.uk/COP/News/Rtcollege_of_paramedics_writes_to_the_rt_hon_matt_hancock_regarding_PPE_for_paramedics.aspx
[7] 'Hospitals turning ambulances away as demand soars, A&E chief says', The Guardian, 25 September 2021; 'Lives at risk from long ambulance waits, say paramedics', BBC News, 11 November 2021.

what is a demanding, exhausting and professionally underappreciated role, paramedics found themselves in the grip of a public health crisis of disastrous proportions. They had to dig deep to find yet another well of energy, resilience, and goodwill. Somehow they found it. Ambulance staff working day and night during the pandemic played a critical role in caring for people in dire need, just as they always have done and continue to do. This book is dedicated to them.

Acknowledgements

I relied on the knowledge, kindness, and generosity of many people in writing this book. Sadly, a large number of them can't be named here because they are the paramedics that I interviewed or observed at work, and whose privacy and anonymity needs to be respected. It is their opinions, experiences, ideas, and stories that make up the empirical material of this study, and I thank them all sincerely for sharing this with me. Learning about paramedic work has been a fascinating and humbling experience, and I hope that this book does some justice to the deep complexities and nuances of that world. In addition, I greatly appreciate the insights of experts in other parts of the ambulance realm that I came to know, such as Ray Bange from Queensland, Australia, and Andy Ellwood from much closer to home.

As for the university world, I must thank a range of people who have influenced this book in particular as well as the formation of my own academic perspectives more generally. Foremost among them are my former colleagues at the University of Manchester with whom I first embarked on research in the NHS, namely Ed Granter, John Hassard, and Paula Hyde. I look back on that time very fondly. While at Manchester—and since—I also learnt a great deal about the subtleties of ethnographic enquiry from such esteemed colleagues as Damian O'Doherty and Dean Pierides. On the specific subjects of uniformed work, street-level bureaucracy, and the huge strains that emergency service work can put on the shoulders of working people, I gained invaluable insights from discussions with current and former PhD students such as Cliff Bacon, Gemma Lord, Fiona Meechan, Jo Mildenhall, and Brian Wierman.

There is a small community of academics exploring the sociology of paramedics and ambulance services, including Rachel Ashworth, Mike Corman, Josh Seim, and Paresh Wankhade, and it's been a great pleasure exchanging ideas and experiences with all of them. Thanks must also go to several participants of the 2021 Professional Services Firms conference at Oxford University who generously commented on my work, in particular Thomas Andersson, Robin Burrow, and Mike Saks. At the University of

York, it has been a privilege working among colleagues in the Work, Management and Organization group, who have been supportive colleagues during a desperately tough time for universities, academics, and students. Special thanks must go to Bill Cooke, Chris Corker, Jon Fanning, Phil Garnett, Lindsay Hamilton, Shane Hamilton, Carolyn Hunter, Ian Kirkpatrick, Simon Mollan, Daniel Muzio, Jane Suter, Simon Sweeney, Kevin Tennent, and Des Williamson. For two years I served alongside Jane Suter as the joint head of this group which, over time, developed a 'span of control' so broad it would have given Lyndall Urwick and V. A. Graicunas real cause for concern. Without Jane's collegiality, support, and incredible attention to detail, I would have drowned under the demands of that management role, and I thank her sincerely.

Jenny King at Oxford University Press has been a very helpful and patient editor, and I would like to thank her and OUP for the support I've had over the years it has taken for me to finish this project.

Finally, I would like to thank my wife Kate Reed for her love and kindness. Not only is she a wonderful soulmate and companion, she is a truly outstanding sociologist of health and illness, and her influence is present on every page of this text. Thank you, Kate, for everything.

Contents

1

A New Profession

'Few callings deserve greater respect than those involving public service.'

Lipsky (2010 [1980]: xiii)

'A Coach and Horses'

22 February 1990. Elephant and Castle, South London. Something important is going on inside the offices and corridors of Alexander Fleming House and its surrounding buildings. A major work of the legendary modernist architect Ernö Goldfinger, this sprawling office complex was envisaged as a giant, interconnected nerve centre of centralized government efficiency. Constructed out of five reinforced concrete buildings connected by walkways, it was, according to design guru Stephen Bayley in a short BBC film, 'a rare example of constructivism' that 'would look more at home in Moscow or Leningrad'.[1] Completed in 1966 it housed the offices of the Ministry of Health and its successor, the Department for Health and Social Security.

By the late 1980s this modernist masterpiece had, like government and public administration in general, declined into an ugly, neglected mess. 'People are funnelled into open sewers' suggested Bayley, 'while traffic congeals on the surface.' Modernist dreams of efficient, rational, technocratic planning had evaporated as centralized ministries unravelled into smaller departments and semi-privatized business units under a series of neoliberal governments. But, back in February 1990, in one of the final events of its life as a government building, events were taking place that hinted at a new beginning amid the general decay of late Thatcherism.

Befitting the technocratic, centralized government policy which was still clinging to life in that era, employment conditions of NHS (National Health

---

[1] The BBC TV programme was entitled *Building Sights* and was originally transmitted on 23 November 1988, https://www.bbc.co.uk/iplayer/episode/p01rqpwx/building-sights-series-1-5-alexander-fleming-house

The Paramedic at Work. Leo McCann, Oxford University Press.
© Leo McCann (2022). DOI: 10.1093/oso/9780198816362.003.0001

Service) staff were negotiated by the Whitley Councils of 'staff side' and 'management side' collective bargaining. For one particular group of NHS workers, in the autumn and winter of 1989–90, this machinery had broken down. Ambulance workers were in dispute with their employers over the 1989 pay offer. The 1989–90 ambulance dispute was about to escalate into 'the most important industrial conflict since the 1984–5 miners' strike' (Kerr and Sachdev, 1992: 127). Starved of funding and recognition for decades and with morale desperately low, ambulance staff represented by five trade unions across the UK took industrial action to defend their pay and conditions from continual erosion, aiming at parity with police and firefighters.

Industrial action took several forms, including overtime bans, working-to-rule, and boycotts of paperwork and non-emergency patient transports. Management responded with lockouts and suspensions. Ambulance workers taking part in the dispute technically were not on strike; rather they were taking part in 'action short of strike' (Clement, 2010: 12–13). Ambulance workers ran their own unofficial 999 response system, continuing to respond to emergency calls while going unpaid, despite Conservative politicians routinely criticizing the unions for endangering the public (Kerr and Sachdev, 1992: 130). The government resorted to calling out the armed forces in some cities, deploying soldiers to emergency calls in olive-green military fire appliances and ambulances. Police officers, firefighters, and St John's Ambulance volunteers were also drafted in to perform emergency medical response (Fellows and Harris, 2019: 34).

Ambulance staff collected donations from the public by shaking buckets outside ambulance stations and in town centres (Conaghan, 2010: 42). Decades before the internet and social media, a petition calling for the government to provide a new pay deal for ambulance staff gained over five million signatures. The dispute was finally ended at meetings brokered by ACAS, including an all-night negotiation session at Elephant and Castle (Clement, 2010: 18–19). In a signifier of the industrial, working-class roots of ambulance work, the staff side sent out for fish and chips during the marathon 20-hour talks. According to one of the unionists 'that was a good sign, as getting to the fish and chip stage reveals progress' (Conaghan, 2010: 84).

On 23 February the lead union negotiator Roger Poole emerged to announce a deal that would end the dispute. He used a phrase that many in the British union movement will recognize: 'Today we have driven a coach and horses through the Conservative government's pay policy' (Conaghan 2010: 85).[2]

His coach and horses metaphor might have several meanings. Clearly Poole is summoning images of solidarity; a well-organized union campaign with wide public support had outmanoeuvred a fiercely anti-union government to punch a hole in its credibility and secure a good outcome for its members. It also resonates with the mobility and drama of ambulance work itself, even to its Napoleonic roots in battlefield extraction; its urgency, its ability to react quickly, to cut through the fog and friction of war and get the job done.

Archived TV news reports of the dispute provide sights and sounds of an era that is distant yet uncannily familiar. Union militancy. Cardboard placards asking motorists to 'Honk Your Support'. Troops on the streets. The vehicles look comically primitive; British Leyland ambulances in white paintwork with two little blue lights on the roof. 'Green Goddess' military fire appliances nose their way out of run-down garages. We hear uniformed, unionized, disgruntled ambulance workers with strong working-class, regional accents. We see control rooms using paper-based dispatch systems and sending emergency calls to ambulance stations via beige-coloured British Telecom telephones. Debates in the House of Commons featured Conservative MPs alluding to 'The Winter of Discontent' when unions 'allowed the dead to remain unburied'. A Labour MP for Islington North by the name of Jeremy Corbyn accused the Secretary of State for Health of 'undertaking a massive conspiracy'.[3]

The resolution of industrial disputes typically sees the various parties claiming successful outcomes. The settlement announced by Poole can be read variously as a pay award of 22.5% or 13.3% (Kerry and Sachdev, 1992: 138). Crews in many regions were dissatisfied and would have stayed in dispute for longer. Amid all the claims and counterclaims, there was a more significant detail buried somewhere in the paperwork—increased pay allowances for 'staff with various paramedic skills' (Kerr and Sachdev, 1992: 138). The dispute had exposed how run-down, overstretched, and clinically

---

[2] 'When ambulance workers drove a coach and horses through government pay policy', workers.org.uk, October 2010; see also: 'Roger Poole obituary', *The Guardian,* 15 July 2015; 'Roger Poole: Trade union official whose impeccable public persona helped ambulance workers defeat the Tory government', *Independent,* 28 July 2015.
[3] *Hansard,* HC Deb, 7 December 1989. Vol. 159, cc852-5, http://hansard.millbanksystems.com/commons/1989/nov/07/ambulance-service-pay-dispute

inadequate the ambulance service had become. Crucially, this 'skills' part of the settlement opened up the possibility of developing more clinically advanced roles for ambulance crews as part of a broader augmentation of the ambulance service's clinical ability. This settlement—now almost lost in the mists of time—turned out to be an important stage for the development of the paramedic role in the UK; it marked the first time that ambulance staff would be paid for having 'paramedic skills' (Fellows and Harris, 2019: 36–7). There had been some pioneering developments in upskilling ambulance responders in various parts of Britain since the late 1960s (Bell, 2009: 254; Fellows, 2020), but the endgame to the 1989–'90 dispute provided a possibility to broaden and embed these, and other, changes as a matter of national policy. Ambulance services and ambulance work could now begin to move from a blue-collar occupation rooted in manual work, first aid, and transportation, to a clinically advanced profession, staffed by highly trained and increasingly autonomous experts.

This book is about the work of paramedics. It is a sociology of a new profession. It explains, describes, and accounts for the work, behaviour, and culture of a group of expert workers who engage in a particular kind of healthcare work. This form of work—much like other uniformed, emergency work such as policing and firefighting—holds an enduring fascination among the public. There is a mystique, even a romance to the work: speeding vehicles, expert technicians functioning under sometimes extreme pressure, caring professionals providing a compassionate service to the public, the exhilaration of rescue, and the tragedy of loss. 'Reality' TV shows and blog-based books about paramedics and ambulance services have been popular hits (book examples include Brent, 2010; Farnworth, 2020; Gray, 2007; Jones, 2020; Pringle, 2009; Reynolds, 2009, 2010; Walder, 2020; Wharton, 2017). While the memoirs tend to be self-deprecating, the TV shows play to well-established public perceptions of ambulance services and paramedics, focussing strongly on the drama associated with life-threatening emergencies; blue-light vehicles tearing through the streets, heroic first responders stoically enduring non-stop trauma, and skilfully intervening to save lives in hostile environments.

Heroic and action-packed portrayals of this kind do not convey a detailed, balanced, and accurate picture of an occupation or profession. Beyond a relatively small sociological literature (Boyle and Healey 2003; Corman, 2017a, 2017b; Granter et al., 2019; Mannon, 1992; McCann et al., 2013; Metz 1981; Nelson and Barley, 1997; Nurok and Henckes, 2009; Seim, 2020; Tangherlini, 1998, 2000; Wankhede, 2012) little is known about the sociology of ambulance, EMT (Emergency Medical Technician) or paramedic work. There is strong public interest, but weak public understanding of even some of the most basic elements of what they do, how they do it, and why. Most people are fortunate enough not to have to call an ambulance often, or ever. The role of the paramedic is fascinating and alluring yet shrouded in mystery.

This book attempts to cut through that shroud. It explores paramedic work in the distinct employment and operating context of the National Health Service in England. The text is oriented around several conceptual themes, but its central focus is the complex dynamics of the transition of this 'occupation' into a 'profession'. Personal interviews and hours of observations will depict the working lives of paramedics in close-up, vivid detail. The book focuses closely on paramedics' own interpretations of the intricacies of their work: their own senses of identity and duty in relation to the public, and the complex and often problematic ways in which they are 'managed' by the NHS trusts that employ them. My account of their views and behaviours will take the reader into some of the backstage areas that are not made available in 'reality' TV and don't feature in the 'overstretched heroes' trope. Overall, it will explore the progress the profession has made in transitioning from a very basic service rooted in transportation and first aid into becoming what it aims to be today and in the future: a clinically focused, autonomous provider of expert health services.

An important element that will regularly recur in this study is the fact that not everything paramedics do (in fact, for most responders, not very much of what they do) corresponds to the image that much of the public holds—that of ambulance work being primarily about rushing to the scenes of life-threatening emergencies, performing drastic and intrusive clinical procedures, before setting off again at high speed to take the stricken patient to hospital. Of course, all experienced paramedics have attended dozens of life-threatening callouts; heart attacks, road traffic accidents, falls from height, suicides, and all manner of other serious and often distressing scenes. But ambulance services are increasingly moving away from their traditional role as solely or primarily focused around responding to life-threatening emergencies such as serious trauma or respiratory arrest; today an estimated 8% of their calls tend to be of this nature.[4]

Health economies are undergoing change. NHS ambulance trusts have expanded in size and remit and their callout activity has grown very

---

[4] 'Fewer ambulance 999 calls to be classed as 'life-threatening', *BBC News*, 13 July 2017.

substantially. Much of their expanded activity has been of the lower acuity kind. Ambulances regularly attend calls which are effectively episodes of unplanned primary care, or psychiatric and social care. While paramedics have always handled calls of this nature (Seim, 2020), in recent years this drift into unplanned primary care has become more directed and deliberate, forming part of an official government policy (Newton et al., 2020). Acute, social care and psychiatric care provision has been at over-capacity for decades as inpatient bed numbers have steadily declined. This has put increasing pressure to keep patients out of overcrowded hospitals. General Practitioners (GPs) are less available for home visits then they once were. Paramedics' extension into these parts of the health system is a result of two broader forces. On the one hand, ambulance trusts have been forced into working in these areas as the emergency response system is relied on by patients to plug shortages. On the other, the shortages have presented an opportunity for government policymakers, ambulance senior managers, and parts of the paramedics' own professional body and leadership to redevelop the paramedic role and the ambulance service. While problematic in many ways, a famous NHS report from 2005 announced a new strategy for ambulances of 'taking healthcare to the patient' (DH, 2005). Not every ambulance callout should result in transportation to hospital—in fact, the aim of this and many other subsequent reports was to reduce the 'conveyance rate' by upskilling and empowering paramedics to treat patients and leave them at scene or at home. When this works well, it has potentially many benefits: the patient avoids an unnecessary and usually unwanted hospital stay, the hospital sees a reduction in overcrowding, and the paramedic crew is more quickly available to respond to the next call—all at lower operating costs. Paramedic services around the world are undergoing similar change (Furness et al., 2020). Services have evolved from what responders used to call 'scoop and run' or 'hump and dump', to something much broader and more complex. Changes to ambulance services' operating models can mean subtle and important adaptation of the work of paramedics, with interesting implications for the standing of the paramedic profession.

While the paramedic is widely regarded as an emergency role, it is probably more accurate to think about it as primarily a healthcare role. Legally speaking, UK paramedics are not an 'emergency' occupation in the same sense as police officers or firefighters. The paramedic is technically an Allied Health Professional, and the vast majority are employed by NHS ambulance trusts on the same pay spine as NHS nurses, physiotherapists, midwives, managers, and other associated clinical staff. Their pay and conditions are negotiated by the public sector trade union Unison, and by the general unions GMB and Unite. The day-to-day role of many ambulance staff 'on the road' is fundamentally not so much about traffic accidents, drug overdoses, suicides, and street violence (although all of these aspects do indeed feature). It is mostly about care and compassion. Their role is to provide treatment, advice, and assistance to patients, support and reassurance for family members and bystanders, and advocacy for patients as a first point of contact assisting them through the labyrinthine NHS bureaucracy. They make differential diagnoses and autonomous clinical decisions—including decisions to effectively discharge a patient by not taking them onwards to a hospital, and to assume the medico-legal risk that these decisions entail. The paramedic is a healthcare profession of increasingly skill, autonomy, and clinical discretion (Eaton, 2019).

In addition, as with any other profession, many paramedics also have supervisory and managerial responsibilities within their employing ambulance trust. Some are involved in collaborating with other services, such as fire and rescue, police, and local and central government in planning around 'preparedness' and 'resilience' for such events as floods, pandemics, and terrorist attacks. Many of them teach, in a practical sense when accompanied by student observers in the field, and in a more academic sense by teaching in university classrooms, where most of the training and education of new paramedics in Britain now takes place. Some paramedics contribute to the continuing development of the profession: lobbying, researching, writing, organizing. Many are highly active on social media. The paramedic world is diverse and complex, and those employed in it are now engaged in all manner of work roles and responsibilities.

Changes in the ambulance world over the last 30 years have been very considerable, accelerating significantly in the last ten years. Back in the late 1980s, English ambulance responders were largely known as 'ambulancemen'. The men wore navy-blue uniforms featuring peaked caps, sky-blue shirts, and ties and blazers with polished buttons. The small number of women ambulance staff wore an archaic uniform featuring ankle-length skirts, tights and driving gloves; an outfit totally impractical for street work. Ambulance responders had a narrow scope of practice, with the role largely restricted to first aid then safe transportation of patients to 'definitive care' which was only available at hospitals. There was little clinical education beyond the ten weeks of training set by the Millar reports of 1966 which represented the UK government's first attempt to set ambulance service standards (Kilner, 2004; Fellows and Harris, 2019: 31). Employment was male dominated. The roots of the occupation lay in transportation, rather than in medical care, with ambulance services across Britain originally provided by local councils before the 1974 NHS reorganization. For a time, some ambulances would be stationed in bus garages. Up until the early 1990s very few units carried defibrillators, with these devices being special provisions funded by charitable donation. Training in the more advanced forms of life-support and cardiac care were in their infancy. Records of parliamentary debates reveal that there were just eight ambulance staff trained to paramedic level in the whole of London in January 1990. Secretary of State for Health Kenneth Clarke, MP once notoriously described ambulance crews as 'professional drivers, a worthwhile job – but not an exceptional one'.[5]

The picture today has 'developed beyond all recognition' (Harris, 2014: 36). 'Paramedic' is now a legally protected title. There is a professional association, the College of Paramedics, with around 10,000 members. According to the classic sociology of professions literature, it now has pretty much all the 'traits' of a profession, such as ethical codes of conduct, peer management, occupational closure, and its own research journals (Leicht and Fennell, 2001: 26; Muzio et al., 2020: 9–10). Paramedics' scope of practice has grown substantially as more and more clinical procedures and protocols are approved for their use in the field (Eaton et al., 2018). Recent official textbooks of clinical practice guidelines run to over 500 pages (AACE, 2013), and most ambulances carry a very wide range of equipment, including more than 40 specific drugs that paramedics are authorized to use autonomously on scene. Advanced roles have been developed in the last decade or so, such as Paramedic Practitioner (the role is designed to work mostly with primary care cases), or Critical Care Paramedic (aimed more at serious trauma and advanced life support). Training has largely moved out of the ambulance service and into higher education institutions. Registration of new paramedics now takes place almost exclusively through one route: students graduating from three-year university degrees in Paramedic Science or a related scheme (Fellows and Harris, 2019). Paramedics have a career structure with a degree of headroom. They are starting to be allowed to independently prescribe medicines. Paramedic-related social media[6] is alive with posts claiming how far the profession has come. To a significant

---

[5] *Hansard*, HC Deb, 11 January 1990. Vol 164, cc 1111, http://hansard.millbanksystems.com/commons/1990/jan/11/ambulance-dispute

[6] See, for example, @WeParamedics, @ParamedicsUK, or #team999.

extent, paramedics have started to outgrow the restraints of their traditional quasi-monopoly employer, the NHS ambulance trusts (Eaton, 2019: 1). Newly equipped with degrees and enhanced clinical scope, hundreds of paramedics have left ambulance trusts to work in other areas of the healthcare economy, such as in GP surgeries or NHS walk-in centres, as attested by the hashtag #notallparamedicsweargreen. Pay, conditions, and bargaining have moved on from the First-World War era Whitley Councils to Band 6 on the NHS Agenda for Change salary band (£32,000–£39,000 plus anti-social hours payments). Even the Goldfinger complex—now Grade-II listed and renovated—is today a series of high-end apartments named 'Metro Central Heights'.

Yet amid all this change there remain some important continuities. Not every aspect of ambulance working life has improved with the passage of time. The nature of ambulance work remains very demanding—in fact it is considerably more demanding now than at any other time in its history. Patient demand is often overwhelming. Call volumes today are approximately ten times heavier than they were fifty years ago (Newton, 2019: 71). Public expectations for a fast-responding, clinically advanced service have grown substantially. Ambulance trusts struggle to cover the huge and increasingly complex volume of emergency calls. Try as they might, they often cannot hit very strenuous performance measures and targets. Everyone is under tremendous operational pressure, from senior management down through the various levels of the mid-management and to the frontline responders. NHS ambulance trusts are often quite unhappy workplaces, bedevilled by interpersonal conflict and mistrust. The paramedic is a new profession, but its current condition reflects the ways in which many professions today tend not to enjoy the high levels of status, reward, and freedom afforded to professions in the recent past (Evetts, 2011, 2012; Muzio et al., 2020; Nichols, 2017).

There is widespread public confusion about the paramedic role; some still regard paramedics as 'ambulance drivers' whose role is restricted to transporting the sick to hospital, whereas others refer to any responding ambulance staff as 'the paramedics' and seem think they are almost doctors in green. We talk of 'the ambulance service' but even that is not right, in that emergency ambulance services in England are provided by ten separate, regional NHS trusts, and also by a growing army of private-sector companies and volunteer responders contracted to the NHS trusts in an effort to meet callout demand. The identity of the paramedic is unclear and in flux. It is very difficult to isolate a specific 'unique selling point' (Eaton,

2019: 1). Part of the paramedics' work, employment, and identity carries a clear NHS style, whereas other parts seem industrial, even militaristic, in nature. As well as possessing some obvious similarities with other public servants such as nurses and social workers, paramedics also share some important common ground with other uniformed occupations such as police officers and firefighters (Braedley, 2015; Charman, 2017). As a whole the paramedic workforce is no longer male-dominated, with 2021 figures suggesting 41% of HCPC registered paramedics are female, although there remain certain areas of the pre-hospital field (such as air ambulances) where women are heavily outnumbered by men.[7]

Through close-up qualitative inquiry, the book will demonstrate that the paramedic profession in the UK has made very significant progress from that starting point in the 'coach and horses' dispute. It has gone from being a neglected and desperately underfunded part of the NHS, almost forgotten in the byzantine DHSS bureaucracy and its Whitley councils, to an increasingly important Allied Health Profession that continues to grow in stature, clinical scope, and political and educational influence.

The book will also show, however, that despite obvious progress, the pre-hospital ambulance field remains a very difficult working environment for any worker to occupy, 'professional' or otherwise. The work can be extremely demanding and exhausting. It carries chronic risk of serious physical and psychological injury (Granter et al., 2019; Maguire et al., 2014). And, as if the work itself was not difficult enough, numerous official investigations, staff surveys, and media stories also attest to deeply ingrained cultures of conflict and mistrust in ambulance services, sometimes involving weak organizational safeguards against workplace abuse, bullying, and malpractice (Lewis, 2017; Manolchev and Lewis, 2021). Despite the clinical improvements an operator culture of hyper-masculine authority remains in place which (at its worst) tends to encourage bullying behaviours and where admitting to errors, asking for help, or raising concerns are perceived as a weakness or a risk to one's employment (see Brewis and Godfrey, 2019; Tangherlini, 1998).

So, alongside all the positive developments of professionalization, the book will also document the intense difficulties that remain central to working life in this environment, which at times seems a throwback to the industrial relations strife of the 1980s. The ambulance world can be

[7] https://www.hcpc-uk.org/resources/reports/2021/diversity-data-report-2021/ (accessed 21 March 2022).

a hostile environment for professional work. The work is demanding and exhausting. The employment conditions can be poor, and the management culture can be unsupportive. Ambulance services face chronic problems of employee burnout and are experiencing a mental health crisis (Lawn et al., 2020). Growing patient demand continues to outpace the resources available to meet it.[8] Organizational pressure is exacerbated by a blame culture and a workplace environment that can sometimes turn toxic.[9]

Amid all this strain and conflict, however, it is undeniable that the paramedic profession has developed rapidly in recent years, to the point where, in terms of clinical input and discretion, paramedics in the UK are among the most advanced in the world. The job offers a respected social role, the ability to help people in need, high levels of discretion and autonomy, a degree of security, a decent workplace pension, and a career track. That is a lot more than is provided by other forms of employment. Compared to many other forms of work or career, the role of the paramedic has certain powerful advantages. Qualified paramedics are in shortage and their skills are in demand, leading to strong employment opportunities. A degree in paramedic science—quite unlike many other undergraduate degrees—usually leads very quickly to employment in that field. Other service-sector jobs are increasingly precarious, heavily micro-managed and offer little to no job satisfaction, self-realization, or career headroom (Dixon, 2009; Korczynski, 2003; Thomas et al., 2020). Thousands of jobs in skilled occupations and professions are at risk of redundancy due to automation, offshoring, disruptive change, and financial uncertainty (Rafi Khan, 2018; Susskind and Susskind, 2015). But the corporeal, embodied, localized, and interpersonal nature of paramedic work, coupled with its undoubted social value, makes it almost impervious to these particular threats. The job holds real attractions: a paramedic gets to help people, to serve the public, to work outdoors, and to be allowed into interesting backstage areas of life. The job is varied and unpredictable. It is not uncommon for people employed in the emergency services world to describe their role as 'the best job in the world' (Chetkovich 1997: 7).

In-depth reflections on all of this will arrive later in the book. But before that, there is a need to lay out the basics of this unique professional and working environment. The remainder of this chapter provides an

---

[8] A 2020 policy document from NHS Providers estimated a shortfall in funding of £237.5m per year. 'Securing the right support for ambulance services', NHS Providers, November 2020. https://nhsproviders.org/securing-the-right-support-for-ambulance-services

[9] 'Revealed: The hidden crisis in Britain's ambulance services', *The Spectator*, 30 August 2014.

introductory description of basics of the NHS ambulance service in England. How does the ambulance service operate in this country? How is it organized, and how does it put the paramedics to work?

The Organization of England's Ambulance Service

In contrast to many other societies where ambulances and paramedics are employed or contracted by fire departments, the UK's provision of emergency[10] ambulance services is almost entirely the preserve of ambulance trusts that are part of the National Health Service. There is some private, for-profit provision but this is mostly for non-emergency, planned transport to and from hospital, and for standby cover at major public events. There are also some charitable organizations that provide ambulance services, mostly on a small-scale contracted basis, bought in by trusts to help them meet response targets. Air ambulances (helicopters usually crewed by a pilot and two critical care paramedics or doctors) are operated on a regional basis by charities. In some localities (and rather controversially) the Fire and Rescue service will also send resources to certain types of medical call (McCann and Granter, 2019: 223–4). Community First Responders also exist—these are members of the public trained by NHS ambulance trusts, who carry medical equipment in their own private vehicles to provide a first response on scene before an NHS ambulance crew can arrive.

When a person in England calls 999, the call goes first of all to a BT operator, who will ask which service the caller requires. If they reply 'ambulance' then their call will be connected to an Emergency Operations Centre (EOC) operated by one of England's ten NHS ambulance trusts. NHS call handlers will assess the call with the aid of call-prioritizing computer systems such as AMPDS.[11] Call handlers take basic information about the call and work through the decision-tree, leading to the call either moving over to dispatch, or being resolved over the telephone via 'Hear and Treat' protocols. The dispatch side of EOC will allocate the call to a vehicle, and basic details of this call quickly find their way to crews who have 'booked on' to

---

[10] Ambulance trusts also provide a non-emergency transport service for patients to and from planned appointments, a provision that in recent years is now opened to private-sector competition (see Hyde et al., 2016: 95–6).

[11] Advanced Medical Priority Despatch System, a system initially developed in the United States in the late 1970s, which tends to be built around traditional trauma and rapid response models of paramedicine.

a shift, via the Mobile Data Terminal (MDT) system that sits in a module built into ambulance vehicle dashboards, or via a bleep on the portable radios that they take with them whenever on shift but outside the vehicle.

The MDT screen provides basic information about a call, including the patient's age and condition, in a compressed format, such as '54YOM, DIB' (54-year-old male, difficulty in breathing). It also displays addresses and maps, with location detail fed into the vehicle's satnav. Around the edges of the MDT display are a series of input buttons for the crew to acknowledge calls, note arrival on scene, cycle through information, and note 'clear' and ready to respond to the next callout. Vehicles also have digital radio sets and push-button arrays for the operation of emergency lights and sirens, as well as other functions such as control of interior lighting, ventilation, and intercom to the 'patient compartment'. Vehicle designs differ across trusts, with vehicles aging rapidly through near-constant use. Driver cabs as well as the patient treatment and transport areas are covered by CCTV. At the time of writing, one of the most common double-crewed ambulances (DCAs) is an adapted Mercedes-Benz Sprinter, with Skoda Octavias widely used as Solo Response Vehicles (sometimes called RRVs, or Rapid Response Vehicles). The Sprinter features an electric powered tail-lift to load and unload the patient trolley.

The working area of an ambulance can be cramped. Much of the space in the rear is taken up with the patient trolley and with equipment storage. There are numerous storage compartments, designed in such a way that equipment is secure and unlikely to fall or move in transit. Commonly used items such as bandages, steri-strips, lubricant, syringes, and tape are placed in Perspex box-shelves with metal tabs to lock them in place. Latex gloves usually sit in the manufacturers' cardboard boxes, often placed on the dashboard or in wall-mounts for easy access. There will be several pre-packed bags of equipment for crews to take with them to patient scenes. Drugs are typically contained in a locked cabinet. The defibrillator and monitor (usually the Lifepak 15) is located on the wall of the ambulance in a rotatable mount. It is a heavy and bulky piece of equipment that is very often detached and carried to patient scenes. Various stretchers and evacuation chairs are folded up and strapped to the interior or stored behind purpose-built panels in the exterior bodywork and accessed from outside. The patient trolley is the largest item in the back of the vehicle, increasingly a complex and expensive piece of equipment with hydraulic suspension. Around the perimeter of the trolley will be two or three flip-down chairs with seatbelts for crewmembers, family members (and student observers)

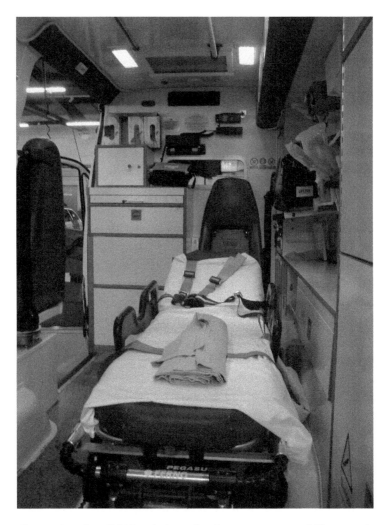

Fig. 1.1 Interior of NHS ambulance 'patient compartment.' Photograph by the author.

to use. A good impression of this highly particular working environment is provided in the image above. I took this photograph with the rear double-doors of an ambulance open, looking from the rear towards the cab. Note the small window between the saloon and the cab which provides (usually) a simple and effective way for the crew to communicate when one is driving and the other is attending a patient.

Paramedics and support workers wear a bottle-green uniform adorned with a series of pockets in a 'combat' style, including belt, epaulettes, and steel-capped black boots. NHS and ambulance trust insignia feature prominently, with some trusts also including a name strip—a first name rather than a surname. The overwhelming proportion of UK paramedics are salaried employees of NHS ambulance trusts, (although there is a growing number of HCPC registered paramedics who work in other NHS settings and a small number employed in police, military, and high-risk industrial settings).[12] Most NHS trusts are large, not-for-profit organizations such as teaching hospitals or mental health providers employing thousands of nurses, doctors, and other allied health professionals. England's ten ambulance trusts are legally the same kind of organization, but their histories and nature differ considerably from hospital or mental health trusts. To begin with, they have traditionally been a 'monoculture', employing staff such as call-handlers, paramedics, and clinical team leaders who typically have little employment experience from outside of the blue light, green uniform world. A hospital trust, on the other hand, will employ people from across dozens of other occupations and employers.

Ambulance service organizations joined the NHS in 1974 meaning they were fairly late additions to the NHS system. Their organizational cultures are a unique mix of NHS influences and a more uniformed, transportation, logistics, and emergency services operating style. Ambulance services have a tradition of being poorly understood and poorly integrated into the rest of the NHS, with opportunities for ambulance workers to share clinical development, research, and training with other NHS occupations and organizations historically quite lacking.

In addition to paramedics, ambulance trusts also employ a range of other frontline ambulance roles mostly analogous to the Emergency Medical Technical (EMT) role that was first developed in the USA. In England, the roles, scope of practice, and seniority of these support roles vary widely, from quite versatile roles with considerable clinical training (such as EMT-1 or EMT-2, which are usually on Agenda for Change Band 4), through to Emergency Care Support Worker (ECSW) or Emergency Care Assistant (ECA) at Bands 3 or 4, with around 6–9 weeks of training. Ambulance trusts also employ hundreds of call handlers and call dispatchers in their EOCs, as well as deploying paramedics or advanced paramedics on clinical support desks, whose role is to advise crews by radio and telephone, resolve 'Hear and Treat' calls, discuss patient needs with other professionals

---

[12] Although some paramedics are not directly employed by trusts, and instead have 'bank' contracts, where can book on to shifts of his or her choosing with one or more Trusts and is paid per shift. Paramedics may do this if they are involved in other forms of work, such as secondment to other parts of the health system, or studying for an advanced degree.

involved in particular patients' care (such as GPs or mental health nurses), and to liaise with hospitals about incoming Accident and Emergency cases. Alongside 'taking healthcare to the patient' the incentive to handle patients outside of hospital also involves the steady growth of 'Hear and Treat' protocols, where a 999 or less urgent 111 call can be handled entirely over the telephone without the need to mobilize a crew. There are also many managerial roles in ambulance trusts—many of them with a paramedic background—from frontline supervisors in the form of Clinical Team Leaders, up to regional heads such as Locality Managers and Area Managers. A large part of their role is operational in nature, such as ensuring that the duty roster is fully staffed so that the trust is best able to meet its response time targets, while also dealing with many complex 'people management' issues, such as employee mental and physical health, absences from work, and investigations and disciplinaries (see Hyde et al., 2016).

Traditionally the route to employment as a paramedic in Britain was via in-service training. Staff joined in trainee roles, starting out on the non-emergency Patient Transport Service or possibly in the EOC as a call-handler. Gaining experience 'on the road' and completing in-house training, they could progress into an EMT role or similar, up into 'intermediate tier' ambulances that mostly respond to less serious cases, then on to a standard DMC (dual-crewed ambulance) emergency vehicle, typically working alongside a qualified paramedic, responding to all kinds of emergency calls. From there, they could progress through the in-service training to gain the IHCD (Institute of Healthcare Development) paramedic qualification, involving 10–12 weeks of in-service training and education including two weeks in an acute hospital setting. Training and education were on-the-job, and highly experiential—a large part of the progression depended on considerations from peers in the service as to how an aspiring paramedic would likely cope with the additional responsibilities. There were informal ways in which people were either encouraged or discouraged from pursuing the various training routes up to paramedic level.

This in-service road to the development of paramedics has undergone drastic change in the last ten years, with the traditional in-house training route to becoming a paramedic now closed.[13] Over the last decade, new paramedics have gained their qualifications through three-year undergraduate degrees. This dramatic change in education and training

---

[13] It has recently become possible for an ambulance employee, such as an EMT, to take a Paramedic Science degree through the new degree apprenticeship route.

is very much in keeping with the paramedics' 'professionalization project' and is not dissimilar to what has occurred in other health service professions, such as Project 2000 for NHS nursing (Brown and Edelmann, 2000). Paramedics in England today are better trained, more flexible, have a higher status, and have a wider scope of practice than at any point in their history. They have also 'discarded' (Kessler et al., 2015) some of their more basic work tasks to other, less qualified staff (such as equipping and checking vehicles). The distinct location and organization of paramedic work, however—out on the streets, focused around the needs of one patient at a time, in vehicles, working solo or with only one other colleague on scene—means that, unlike many other occupations or professions, the discarding of mundane tasks in order to focus on more advanced ones is often very hard to do.

This all means that the paramedic population employed in NHS ambulance trusts is now an interesting mix of persons with different career and education trajectories. There is a core of more experienced paramedics which is typically IHCD-trained alongside a rapidly growing, younger cohort of degree-trained paramedics. Although systems do exist for 'Newly Qualified Paramedics' to operate with in-service supervision and preceptorship, a harsh new fact of life of paramedic professionalization is that degree-trained paramedics are expected to 'hit the ground running' very soon after graduating. A paramedic will be the most clinically responsible person on an ambulance, supervising the work of ECAs and others, and can expect to be the senior clinician present at some very challenging high-acuity emergency scenes at an early age, maybe 22 or 23 years old. This is not only clinically challenging, but also testing for a new paramedic's interpersonal and professional abilities. Many EMTs and related support workers are likely to be highly experienced ambulance staff who have responded to thousands of emergency calls and may have a great deal more practical experience than a newly qualified paramedic.

The most ominous challenge for paramedics and ambulance staff of all kinds is the steady ramping up of demand and operational pressure. NHS ambulance services, like many ambulance services around the world, have for decades faced extremely heavy call volumes, much of it 'clinically unnecessary' if we think of ambulance services in their traditional role as reserved for 'genuine emergencies' (Oshige et al., 2008; Palazzo et al., 1998; Snooks et al., 1998). Part of the reason for ambulance services' crushing demand load is the compound effects emanating from many parts of the overall public services system that are similarly overstretched and struggling. In addition to overcrowded hospitals and oversubscribed GP surgeries, the policing and justice system, the various providers of public housing, social care, and mental health services are all facing high demand, major shortages, organizational challenges, and employee ill-health. Multi-layered scarcity creates interlocking delays and pressures. For example, patients often need urgent care but cannot get a GP appointment. Often they will call 111 (the non-urgent NHS care and advice line) but the problem is insurmountable over the phone, and risk assessment systems will escalate the call to 999 and a possible ambulance callout. Many patients endure long waits for ambulances to arrive. Following crew arrival, patient assessment and onward transportation, there can be further blockages in the system. Acute hospitals in general and their A&E units in particular are chronically overloaded. Ambulance crews often face long waits outside hospitals and in corridors before they can transfer their patients and respond to following calls which start to 'stack' at the EOC. It is not at all uncommon for upwards of 10 or more vehicles to be waiting outside A&E units for hours at a time. This problem is common to both urban and semi-rural settings. A recent news story about Worcester Royal Hospital during the always-difficult Christmas and New Year period mentioned a queue of 23 ambulances waiting outside.[14] Broader estimates have suggested nearly all ambulance trusts regularly missing their mandated performance targets, with a growing number of patients waiting more than an hour for a crew to arrive, and a staggering 500,000 hours of crew time burnt through waiting for hospital handovers.[15]

Numerous systems are in place tracking all of this activity, supposedly to allow managers to monitor and adjust current performance and, in the longer term, to provide auditing and accountability to taxpayers. In keeping with the trends of New Public Management (Bevan and Hood, 2006; Lapsley, 2009), these regimes of governance and their systems of quantitative indictors are often demanding and punitive. Many of these response time targets and clinical standards are oriented around speed of response to calls of various categories. At the time of my fieldwork, calls to 999 deemed to require an ambulance response were categorized as either Red 1—most severe life-threatening emergency, such as cardiac arrest or choking, with an eight-minute response time; Red 2—life-threatening such as stroke or difficulty breathing, also with an eight-minute response time; or Green 1

---

[14] 'Queue of 23 ambulances wait to get into overstretched A&E', *Metro*, 30 December 2019.
[15] 'Ambulance target failures highlight NHS crisis, say health chiefs', *The Guardian*, 30 November 2016.

to 4 (or 'Category C') for any other type of call, such as diabetic problems, suspected fractures, less serious injuries, and falls, with varying response targets set locally, usually between 20 and 90 minutes (Pilbery and Lethbridge, 2015: 7).[16] The targets are linked into complex digital architectures of tracking, measuring, and evaluating (Hyde et al., 2016: 78–9). One of the first things a visitor to a UK ambulance trust HQ will see is an array of KPI data displayed on screens in the entrance lobby. Various numbers, graphs, and indicators display 'real-time' data of the trust's 'performance'. Similar screens have prominent positions around the walls of EOCs. NHS ambulance services float in a sea of metrics (Beer, 2016). Data on calls received, time of dispatch, time to 'wheels rolling', arrival on scene, handover time, and various clinical indicators cycle back and forth around the nodes of the system: front line crews interact with the prompts issued by the MDT in the dashboards of their ambulances, clinical supervisors access the data by logging on to PCs at ambulance stations or by swiping through their smartphones. Managers and accountants at HQ pore over the aggregated details. This digital infrastructure sits awkwardly alongside legacy paper-based systems, and the transition to digital records is far from straightforward.

Performance pressure has stimulated all kinds of organizational responses to try to meet demand and to avoid missing targets or failing audits and inspections. NHS managers devise marketing programmes that advise patients to 'Choose Well'. Changes in the EOC have enhanced the 'Hear and Treat' models, and ambulance crews have been gradually upskilled and given the autonomy to 'Treat and Refer' in the field. Trusts have introduced new vehicles known as Rapid Response Vehicles (RRVs) or Solo Response Vehicles (SRVs)—adapted road cars staffed by one crewmember, with no capacity to transport patients. Their aim is to generate more capacity to hit response times and to enable 'See and Treat' protocols where patients can be seen by a solo responder without the need for

---

[16] The Ambulance Response Programme of 2017 replaced these categories with four broader categories: Category 1 (Life threatening, response target seven minutes); Category 2 (Emergency, 18 minutes); Category 3 (Urgent); and Category 4 (Less Urgent). Categories 3 and 4 have no average response target, but instead have an expectation of a 90th percentile response target of, respectively, 2 hours and 3 hours. https://www.england.nhs.uk/wp-content/uploads/2018/10/ambulance-response-programme-review.pdf

ARP is widely seen as a welcome and much needed change, but one that is unable by itself to resolve the chronic issue of excessive call volume. It helpfully provides EOCs more time to decide how to categorize a call which discourages the prior practice of sending whatever resources happens to be nearest. An interesting and largely positive review of the effects of ARP can be found here, written by a paramedic who felt unable to post their real name for fear of management retribution. https://www.sochealth.co.uk/2017/10/19/ambulance-response-programme-socialist-analysis/

Fig. 1.2   The driver's seat in an NHS ambulance Solo Response Vehicle. Photograph by the author.

transportation. Figure 1.2 shows the driver's view of the interior of an RRV. The high density of IT and communications systems mounted in the cockpit conveys something of the high cognitive load placed on ambulance staff performing emergency driving, especially when working solo.

The body that represents ambulance trust management (the Association of Ambulance Chief Executives) warns of overstretch but runs the risk of appearing ineffective if it complains too much about being overloaded and under-resourced. Instead, it tries to develop a narrative that ambulance trust leadership is sufficiently flexible and responsive to cope with the mind-bending volume and complexity of work that comes in. An interesting illustration of this can be found in an AACE document 'Leading the Way to Care', (AACE, 2015: 6–7) which contains a rather terrifying spider diagram of the huge volume and range of demand scenarios that an NHS ambulance trust is supposed to be willing and ready to handle. Examples include: 'my son has advanced cancer, he's in extreme pain and struggling to breathe', 'there's been an explosion in a factory and it has collapsed with several people inside', or 'I've run out of inhalers and the pharmacies are all shut.'

The pre-hospital world is a complex field; a varied and confusing mix of bodies. There are regulators such as the Health and Care Professions

Council (HCPC) which regulates registered paramedics and the Care Quality Commission (CQC) which monitors the clinical practice of ambulance trusts. The Association of Ambulance Chief Executives (AACE) is the managers' representative lobby group. The College of Paramedics is the professional body of the paramedics. The trade unions Unison, Unite, and GMB represent thousands of ambulance staff. Clinical protocols for paramedics are developed by a group known as JRCALC—the Joint Royal Colleges Ambulance Liaison Committee. This committee has members from many professions, but is dominated by doctors, such as those representing the Royal College of General Practitioners, the Royal College of Anaesthetists, and the Royal College of Emergency Medicine.[17] While paramedics have a seat on this table, they are some way from being in charge of the clinical direction and scope of their own profession (see McCann et al., 2013). Getting to Band 6 has been a long battle, with the government and some of the other NHS professions not always supportive.

As part of the NHS, the UK ambulance service is 'free at the point of use'—no patient is ever charged for calling or using the service. Paramedics and other ambulance roles—like the professions of nursing and medicine—typically score very well in public opinion polls as being highly trusted professions. While many recognize the severe demand pressures on the service and complaints about slow response times are common, the general public tends to agree that any blame for this lies not with individual responders, but rather with the government for underfunding NHS trusts. In fact it is quite common to read public comments describing ambulance staff as 'heroes' or even 'angels in green'.[18] 'Reality' TV programmes such as *Ambulance, Helicopter Heroes,* and *999: What's Your Emergency?* play to this 'hero' narrative, reporting not just the emotional and physical intensity of the work itself, but sometimes also on the extreme pressures of organizational demand (Granter et al., 2015). Various ambulance-related organizations are becoming increasingly sophisticated with their conventional and social media outputs in order to bolster public trust and better explain what they do.

On the other hand, ambulance services also regularly receive bad press. There have been numerous official investigations into ambulance delays and care failures (National Audit Office, 2017; House of Commons Public

---

[17] https://www.jrcalc.org.uk/committees/main-committee/ (accessed 20 April 2020).
[18] 'Angels in green' was a comment from a patient to the BBC. 'A day in the life of the ambulance service', *BBC News*, 30 November 2016; available at: https://www.bbc.co.uk/news/live/health-38144370 (accessed 20 April 2020).

Accounts Select Committee, 2019). News stories about staff burnout are very common,[19] as are media accounts[20] and official investigations of entrenched cultures of bullying and harassment (Lewis, 2017). NHS staff surveys[21] and Unison trade union surveys[22] reveal that sickness absence, grievances, heavy workloads, bullying, violence, and low morale are consistently very serious problems, with ambulance trusts often faring badly in comparison with other types of NHS organization. Tragically, there have been many examples of ambulance staff suicides in recent years (Mars et al., 2020).[23] Paramedics in England are professionals, but they operate in an employment context characterized by severe strain and conflict (McCann et al., 2013; McCann and Granter, 2019). This context often corrodes and restricts the full expression of the professionalism that paramedics have worked so hard to gain.

Understanding *Homo Paramedicus*

Having introduced the history and the operating structure of ambulance services, we can now explore the people who embody and inhabit the paramedic profession. Organizations have cultures, as do occupations and professions. Fields—following Pierre Bourdieu—are often said to possess a 'habitus'; a patterned set of practices, or an 'acquired system of generative schemes', that shape and structure the actions and behaviours of those comprising it (Bourdieu, 1977). Bourdieu famously applied his theory in the classic text *Homo Academicus* (Bourdieu, 1990), discussing the distinct behaviours, identities, reward systems, and structuring hierarchies of university professors. A similar socio-cultural logic can be applied to the action and structure of any sector, occupation, industry, or profession. Several NHS organizations and occupations will have their own versions of, for

---

[19] 'Huge rise in paramedics off sick with stress', *BBC News*, 2 March 2019; 'Crumbling Britain: Bullying, stress, and death in the ambulance services', *The New Statesman*, 2 January 2020; '"We're all approaching burnout": Paramedics on the toll of tackling coronavirus', *The Guardian*, 23 April 2020.
[20] 'Bullying of desperate 999 call handlers 'led to suicide", *Daily Telegraph*, 13 February 2017; 'East of England Ambulance Service staff "silenced" over bullying', *BBC News*, 10 February 2020.
[21] https://www.nhsstaffsurveys.com/Page/1074/Latest-Results/Ambulance-Trusts/ (accessed 24 April 2020).
[22] 'UNISON survey reveals scale of secret stress among ambulance workers'. Unison, 10 April 2015. Available at: https://www.unison.org.uk/news/article/2015/04/unison-survey-reveals-scale-of-secret-stress-among-ambulance-workers/ (accessed 24 April 2020).
[23] National Institute of Health Research, https://evidence.nihr.ac.uk/alert/ambulance-staff-who-respond-to-suicides-need-more-support/ (accessed 23 June 2020).

example, a 'habitus of rescue' when patients are in acute danger (Mackintosh et al., 2014). There is something about NHS work, emergency work, and specifically ambulance work that attracts and retains certain people. A habitus takes different forms across various fields and sub-fields. Bourdieu argues that the habitus both structures and is structured by the actions of the persons in the field. This means that the habitus has the potential to undergo change as the field adapts over time. This is certainly the case in the ambulance world.

The identity and professional bearing of what was once considered a 'typical' paramedic are changing. A caricatured example might look something like this. He is male, IHCD trained, employed for 20+ years in one ambulance trust, and very much 'a product of the system'. He is respected by peers and perhaps slightly feared by junior staff. But he's also tired and burnt out. Would he be a union member? Certainly. A College of Paramedics member? Possibly. His time in service and huge experience marks him as a respected paramedic on station and out on the streets. But some colleagues are cautious of his 'old school' attitude. He's part of a 'boys club' on station. He 'enjoys a good moan'. His greying hair is tightly cropped, but the green uniform doesn't quite obscure his somewhat rotund trunk. Highly critical of ambulance managers and the employment hierarchy, he is nonetheless devoted to his patients, at least to those he considers deserving. Offstage, he will happily engage in banter about how so many of the 'punters' are timewasters whose calls never really warrant an ambulance response. Trained for an outlook focused mostly around life-saving procedures for the critically ill and injured, he can be reluctant to use painkillers. He refers to emergency calls as 'jobs' and is particularly critical of 'social' or 'psychiatric' jobs. The adjective often used to describe his outlook would be 'cynical'.

Such a caricature would be in keeping with the traditional understanding of an 'ambulanceman' as documented in the early studies of paramedic and EMS work (Palmer, 1983). It corresponds to Donald Metz's classic characterization of ambulance responders as 'blue-collar professionals' (Metz, 1981: 59–73), and of EMS's roots as a 'trade' rather than a profession.[24] It is also similar to the traditional, action oriented, and highly masculine habitus of the 'crewroom culture', associated with fire services or the police

---

[24] 'Is EMS a Trade or a Profession?', *Journal of Emergency Medical Services*, 28 July 2016 (accessed 25 January 2022).

(Bittner, 1967; Chetkovich, 1997; Charman, 2017; Desmond, 2007; Loftus, 2012; Reiner, 2010; Scott and Tracy, 2007).

But this persona is changing rapidly. The 'professional' is becoming more prominent and the 'blue-collar' aspect is dwindling (McCann and Granter, 2019). Ambulance services (like other emergency services) are becoming increasingly sophisticated and are being relied on ever more heavily by the public and by government. The entire health economy is changing and with it, the roles, identity, training, education, behaviour and status of paramedics and other healthcare workers (Eaton, 2019; Saks, 2021). Ambulance services are becoming more gender-equal in their makeup. The trade unions are less powerful and the professional association more influential. Technologies, equipment, vehicles, drugs, treatments and protocols have changed out of recognition from the time of Metz's study, in which one character claimed that 'first aid and common sense is all it is' (Metz, 1981: 71). Changes are enveloping ambulance services with the enlargement of paramedic discretion and education probably representing the most important change. Yet NHS ambulance trusts still feature organizational, operational, and cultural problems to the point where low levels of trust, a blame culture, and widespread cynicism, conflict, and exhaustion remain common workplace experiences.

A major source of tension lies with the way in which paramedics are employed. Although the paramedics are growing in sophistication, flexibility, and competence, they are often still employed in a restrictive and mechanistic fashion by NHS trusts which prioritize the short-term and narrow needs of response time targets (Newton et al., 2020). Training, professional development, research, and service to the general profession all take a back seat while the trusts emphasize 100% roster utilization and meeting the mandated percentage of Category 1 and 2 calls within the response time standards.

However, while the managerial and organizational goals of the employers often pose acute challenges to the expansion of paramedic professionalism, there is one aspect of the UK's monopolistic NHS trust structure that has played a positive role. British paramedics have arguably benefitted from the centralized nature of paramedic employment and health policy. This has allowed the College of Paramedics to extend its influence into approving universities' educational curricula, helping to enlarge the scope of practice and take paramedic skills and paramedic standards out of market competition. While the development of the paramedic profession rose from a very low base in the 1980s (Harris and Fellows, 2019), the paramedic profession in the NHS has recently progressed quite rapidly and is arguably in one of the strongest positions in the Anglophone world. Although significant early development of the paramedic role took place in North America (Bell, 2009), further progress in the USA in recent years has been hampered by the radical fragmentation of employers, skills and certifications. EMS provision in the USA and Canada is a very complicated patchwork of different types of organization: from fire departments to private equity operators (Bell, 2009; Corman, 2017b; Seim, 2020). Employment conditions can be particularly poor in the States, where EMT and paramedic pay is significantly below that of police officers, nurses, and firefighters. The USA has thousands of organizations providing EMS of some kind (many of them privatized, and many of them voluntary). The Bureau of Labor Statistics records the 2019 median pay of an EMT or paramedic at $35,400 per year, or just $17.02 per hour.[25]

There is now a detailed and extensive paramedic education and training literature (see, for example, Fellows and Fellows, 2012; Harris and Blaber, 2014; Pilbery and Lethbridge, 2015). Understandably, these texts focus on the technical side of the work. But in doing so they tend to downplay or sideline employment and welfare issues, typically providing little consideration of the difficult employment and management culture into which paramedics sit. Moreover, they tend to construct professionalism in a rather uncritical fashion. The responsibility of being 'professional' is loaded onto the shoulders of the individual practitioner. Rather like the broader notion of 'employability', professionalism is portrayed as a one-way street in which managers' (and patients') responsibilities in creating an appropriate and supportive environment for professionals and professionalism are rarely considered. In such a context, 'professionalism' risks being something not owned and defined from below, but imposed from above (Evetts, 2011). This can render 'professionalism' as something largely meaningless, or—worse—something coercive, defined by the employer rather than the professionals and their representative body. Professionalism can easily be attenuated so that its meaning is restricted only to employer-defined and employer-friendly behaviours represented and controlled by the structures of managerialism, audit, and compliance. Professionalism can also be cast in an excessively consumer-centric way, where a model professional conforms to the market-driven notions that the 'customer' is always right.

---

[25] US Bureau of Labor Statistics, https://www.bls.gov/ooh/healthcare/emts-and-paramedics.htm (accessed 27 April 2020).

One of the key purposes of the book, therefore, is to unpack and explore the complex and contested meanings of professionalism. The identity, expertise, discretion, and licence associated with professionalism are powerful and sought-after. The licence to operate on a discretionary and largely self-directed basis as 'a professional' is not something easily established. It is also influenced and restricted in important ways by broader contexts of consumerism and the employment relationship. Professional status can also be abused. It carries great responsibility. It is also easily lost. And professionalism isn't only about status. It is a form of behaviour and service whose realization requires effort and cooperation from several sources, especially in the troublesome context of street-level work. I aim to portray paramedic professionalism as something always in development, contextualized in a difficult employment context, and as something meaningful, precious, but potentially at risk.

The Story to Come

The rest of this chapter will briefly describe the contents of the book. The following chapter will place the paramedic profession into a broader conceptual discussion on the nature of professions, professionalism, and professional work. There is a long history of sociological writings on professionalism, with large parts of the debate formed in the 1960s, a time where today's very familiar ideas of a 'knowledge economy' and a service-based rather than industrial-based society were first taking shape. That time also reflected some important early forms of change in gender relations and a vast increase in educational opportunities. Since then, the definition of 'profession' has undergone substantial change. In the early days of the sociology of professions literature, it tended to refer to what now would be considered elite professions, such as architecture, accounting, medicine, and law. Recent decades have witnessed a huge growth in occupations seeking 'professionalization', with many of them achieving significant successes. Chapter 2 will, therefore, explore the professional claims of paramedics, the areas of their expertise, as well as the areas where they still struggle for legitimacy and reward. It will do this via a discussion of a range of sociological theories about professionalism, professional work, and technical work (Abbott, 1988; Barley and Orr, 1997; Freidson, 2001; Leicht and Fennell, 2001; Orr, 2006), as well as some important notions that have emerged from public administration, notably the concept of 'street-level bureaucracy' (Lipsky, 2010), and the challenges of New Public Management and managerialism (Evans and Harris, 2004; Evans, 2010; Exworthy and Halford, 1999). It will ground the discussion of paramedics in a broader discussion of the 'state of the art' of the sociology of other emergency and uniformed occupations (Bacon, 2019; Charman, 2017; Desmond, 2007; Loftus, 2009), and will also review the small but powerful sociology of ambulance work, something that has been in abeyance for some time, but now seems to be making a welcome comeback (Corman, 2017b; Kyed, 2019, 2020; Seim, 2017, 2020).

Chapters 3–6 provide the book's main contribution. Here the book draws on the experiences, opinions, and stories that are the lifeblood of the ambulance world. The first of these four empirical chapters provides a broad overview of the structure, duties, and rhythms of paramedic work, aiming to immerse the reader into this field, before the following chapters explore more specific elements of the work and the occupation in more depth. Chapter 4 documents the paramedic operating culture. We explore the physical, mental, and emotional stresses and strains of the work in Chapter 5. Chapter 6 concludes the empirical material by exploring the various meanings of professionalism, both formal and informal. Throughout these empirical chapters, the aim is to demonstrate, using as much in-depth illustration as possible, the human side of ambulance operations; the opinions, views, and behaviours of those involved in this complex work, day after day. Paramedics' reflections range over a very wide array of the human condition, including philosophical reflections on the randomness of life, the curse of illness, the pain and absurdity of accidents, and the terrible sadness of death and loss. They also feature incessant griping, moaning, joking, and gossip. I interpret the paramedics' words and deeds as contributions to both the understanding and the constructing of their profession. I discuss professionalism not just as a form of social status that a person and a group can hold, but also as something less permanent, less tangible, and not always achievable; something potentially present in the moment as a form of action. This is something I call 'professionalism in process'. It occurs when ambulance work progresses successfully and appropriately in the field, and when the public is as well-served as it can be by a professional ambulance response. Ultimately I argue that the paramedics can be conceptualized as 'street-level professionals'—a concept I develop out of the classic work of Lipsky (2010/1980) and Metz (1981). These chapters are followed by a short conclusion and by an appendix that provides a brief description of the research process I adopted as I developed this book, explaining the research methods I used and the various challenges and complexities I faced; practical, empirical, interpretative, and ethical.

In my time with the paramedics, I came to regard the experience as somewhat schizophrenic. For someone totally unaccustomed to their style of work, I found it always fascinating and occasionally exciting. Accompanying them as an observer on shifts was an extremely rich sociological experience. I learnt an enormous amount and I greatly enjoyed their company. I felt privileged to be allowed into a small corner of their world, if only for a short time. But I also found the experience physically and emotionally exhausting. At times I felt fear, anxiety, and great uncertainty. I could appreciate the enormous attractions of the job. Working out on the road and not knowing what would come up next is fascinating: the ambulance worker continually meets new people, experiences strange and uncommon events, and works out ways to improvise solutions to particular problems. I could appreciate the freedom offered by operating out in the field with minimal supervision. I witnessed the uplifting experience of serving the public with compassion and dignity. I also saw all manner of things that were irksome and difficult. I saw paramedics frustrated with inappropriate callouts and their exasperation of being forced into unplanned overtime. I came across staff expressing their annoyance with restrictive and sometimes petty forms of managerial control, and grappling with stultifying paperwork and clunky IT systems. I guess that's why they call it 'work'.

But there was something very particular about work of this kind. The experiential peaks and troughs of paramedic work can be polarized to an extreme degree. Paramedics experience the ecstatic highs of saving a life or delivering a baby, the dreadful lows of death, illness, and random accidents, and everything in between. Their behaviour reflected this. I saw examples of kindness, care, and compassion towards patients, family members, and colleagues. I also heard sharply critical and cynical comments in sealed-off, crewroom-only spaces. Ambulance staff have to be generalists— they need to be comfortable with any kind of patient they are called to. Their work encompasses an extraordinarily broad range and often carries a deep cognitive load (such as emergency driving, or handling very complex, life-threatening acute cases). They need a range of clinical abilities and typically they possess excellent communication skills with which to interact with patients, family, and the general public. There is the expert, technical side of pre-hospital medicine, such as the complexities of interpreting ECG outputs, or working out an accurate differential diagnosis. Then there is also the 'operational' side of the work, its almost military bearing; green uniforms and heavy boots, remote communication systems, powerful vehicles, the great outdoors, grimy and run-down garages, stations, and standby posts. Like many other large organizations, there is also a huge digital infrastructure that sits awkwardly in and around the human behaviour, to an extent trying to control, coordinate, and evaluate all this activity; a virtual cloud of digital analytics, big data, and 'business intelligence' regarding targets, response times, quality indicators, healthcare outcomes, and staff performance management (see also Corman, 2017b). Some of this is beginning to be automated—certain calls are dispatched through to ambulance vehicles automatically, with the aim of shaving seconds off response times. There is inter-service collaboration with police and fire, but also rivalries, jealousies, and conflict. Ambulance staff complain that the Fire and Rescue service is underutilized and they jokingly describe firefighters as 'water fairies'. Paramedics regard certain NHS facilities and professions with respect and gratitude. Others are portrayed as obstructive or unfriendly. Overall, however, the emerging story of the paramedic as a new profession is one of substantial change and clinical progress. There remain reasons for optimism. Life as a paramedic can be extremely rewarding. For all its strain, conflict, risk, and overload, the work of the paramedic is imbued with distinct forms of meaning and dignity that are difficult to find in a contemporary world so disfigured by the atomizing, selfish, and distrustful imperatives of neoliberalism.

The concluding chapter will reflect on the book's purpose and overall argument. It will discuss the broader meaning of the development of a new profession—what 'a profession' and 'professionalism' might provide for an occupation under such uncertain conditions. It will offer some final thoughts about the ever-changing nature of professions and occupations in a contradictory world in which expert knowledge is ever more valued and respected, yet simultaneously undermined and threatened.

2

Conceptualizing Emergency Professions

'Front-line workers develop a set of shared, unwritten rules that consciously and often unconsciously drive their work behaviors.'
(Riccucci, 2005: 5)

Paramedics do not sit comfortably into any pre-existing category of occupation or profession. New professions—like new technologies or new business models—can stimulate social and organizational change, often forcing other systems and groups to adapt in response. To borrow a contemporary 'business and management' phrase, paramedics are 'disruptive innovators' (Newton, 2019: 74) in the medical field. They have recently gained increased technical competence at low cost in comparison to many other clinical professions. Various groups governing, managing, and reforming health economies have been intrigued by the enticing opportunities provided by a more advanced and versatile ambulance service, all the while being rather unsure what exactly paramedics are and what exactly to do with them. Other groups, occupations, or professions are wary of the intrusion of paramedics into 'their' spheres of competence and are less supportive of changes that might lead in those directions.

Paramedics in England have a distinct history and background, and the multiple fields and jurisdictions in which they work and interact (pre-hospital emergency medicine, unplanned primary care, psychiatric care, fire and rescue response, disaster management) have also undergone significant change in the last 30 years (Fellows and Harris, 2019; Newton, 2019; McCann and Granter, 2019). The paramedic field is no longer a monoculture—its own occupational remit has expanded and diverged as new specialisms and employment opportunities have arisen in recent years (Eaton et al., 2019; Eaton, 2019). Academic research on professions and occupations is similarly wide-ranging. Literatures and sub-literatures across the disciplines of sociology, anthropology, public administration, social policy, criminology, and management are all of potential value in

The Paramedic at Work. Leo McCann, Oxford University Press.
© Leo McCann (2022). DOI: 10.1093/oso/9780198816362.003.0002

making sense of the new paramedic profession. Selecting concepts and ideas from across the spectrum of scholarly ideas and traditions that might best connect to the paramedic experience is no simple task.

This chapter is my best effort at doing so, laying out the conceptual background for the book to come. It describes and explains five distinct but interrelated scholarly traditions that my analysis draws upon. Five might seem like too many, but the breadth and versatility of the paramedic role lend themselves to a broad scope of analysis. The five conceptual literatures I use are: (1) professionalism and expertise (such as the writings of Abbott, 1988; Etzioni, 1969; Freidson, 1986; Larson, 2013/1977; Wilensky, 1964) and more recent work by Eyal, 2013 and Evetts, 2011); (2) management and organizational control, especially the modes of measurement, standardization, and administration associated with New Public Management (Hood, 1991; Lapsley, 2009); (3) street-level bureaucracy (beginning with Lipsky before its expansion and development by more contemporary writers such as Hupe and Hill, 2019; Kaufman, 2019; Riccucci, 2005; Vinzant and Crothers, 1998; and Zacka, 2017); (4) occupational, technical, uniformed, and care work (Abbott and Meerabeau, 1998; Barley and Nelson, 1997; Loftus, 2010; Orr, 1996, 2006); and finally (5) the notions of 'edgework' and 'extreme work' (Lyng, 1990; Granter et al., 2019). I locate the behaviour, identity, and practices of the emerging paramedic profession as somewhere inside this shifting mass of concepts.

Paramedics possess core traits of both the professions and the technical occupations. They are employed by bureaucratic organizations which exert considerable organizational control, yet the location and nature of paramedics' work give them wide autonomy and compel them to be proactive improvisers. At heart, they are highly practical care professionals; humane, sociable, versatile, and clearly part of a distinct 'occupational community' (van Maanen and Barley, 1984). This ties them strongly to an NHS care ethos. Yet the nature of their job also contains work and identity elements associated with other uniformed occupations that operate in different settings and jurisdictions (such as police officers). While all NHS clinicians face clinical emergencies, they rarely—if ever—do so outside of managed and contained clinical settings. Some of the work activities of paramedics fit squarely into those of uniformed emergency workers, activities that can often be understood in terms of edgework (Lyng, 1990) and extreme work (Granter et al., 2015, 2019). These latter two concepts powerfully account for the paradoxical situation whereby challenging, high-involvement, high-discretionary work and activities can be both mentally and physically hazardous, yet highly seductive and rewarding.

The following sections of this chapter go through these five literature branches in turn. The aim is to arrive at the end of the chapter with a strong conceptual base upon which to place the empirical evidence to come. In doing so, I hope to locate and embed this 'disruptive' new profession into a core of sociological literature, with the aim of helping us to better understand its complex nature.

The Sociology of Professions

There is a very long lineage of sociological and historical writing about the nature and role of professions in contemporary societies (Leicht and Fennell, 2001; Muzio et al., 2020). Among the most prominent is the work of Freidson (1970; 1986), Larson (2013/1977), and Abbott (1988). These texts broadly argue that 'the professions' are high-level occupations that involve advanced education, a body of knowledge, a high degree of self-regulation, and enjoy a special licence that grants a monopoly of practice (Abbott and Meerabeau, 1998: 2–3). The more historically focused accounts (Larson 2013/1977; Perkin, 2002) described how the original professions (clergy, law, and medicine) and medieval guilds were followed by the development of the modern professions of the industrial age: architecture, accounting, engineering, and others that grew in variety, scope, and influence during the nineteenth century. Professions were an indispensable part of the fabric of industrial society. More recently, the long-run economic recovery from the Second World War was partly based on a dramatic expansion of scientific-technical development, in which a range of new and old professions played important roles. Academic observers predicted the coming of an 'expert society' or 'knowledge economy' (Bell, 1976; Kerr, 2001/1963) and considered the possibility for a wide range of occupations to be transformed into professions (Wilensky, 1964).

There is a traditional consensus across the sociology of professions literature about the attributes or 'traits' of a profession. For an occupation to be considered 'a profession' it generally needs to feature all of the following: an intellectual knowledge base mastered by extended university-based education and qualification; socially valuable tasks such as those relating to health, science, engineering, or legal rights; selfless practitioners motivated by altruistic service to clients' and society's needs and to the building of the profession and knowledge base; high degree of operator autonomy; members with a long-term commitment to the profession who enjoy a sense of community; and, finally, well-developed codes of ethics that guide behaviour and shape the profession's values and norms (Goode, 1957: 194; Leicht and Fennell, 2001: 26).

Professionalism is not simply a form of status that is conferred on an occupation following a successful professionalization project. It also carries considerable social expectations around appropriate behaviour, identity, and conduct. There are distinct responsibilities that professional office-holders need to uphold to themselves, to their profession, and to society. Professionalism, then, is not only a form of organizing or a marker of social status. It is also something that is socially constructed and enacted by occupational members' everyday actions, speech, and dispositions (Dingwall, 1977; Maitlis and Christensen, 2014; Maynard-Moody and Musheno, 2003). These actions aim to correspond appropriately to internally and externally generated expectations about the social roles of a profession.

The sociology of professions literature tends to focus its attention on the most high-status, most highly rewarded professions (doctors, architects, lawyers) and to 'professional services firms' (auditors, consulting firms, legal partnerships). Employment in these exclusive roles is possible only via a traditional, general degree from an elite university, and the gaining of subsequent profession-specific credentials upon employment. (The possession of sufficient cultural capital and elite connections also help considerably.) Somewhat neglected in these discussions is the position of occupations of less rarefied status, ones that play important roles in a society but are not part of its elite structure (Nicklich et al., 2020). Occupations such as nurse, IT administrator, or midwife are usually placed in the category of semi-profession, para-profession, or technical occupation (Etzioni, 1969; Freidson, 1986: 48–57; Leicht and Fennell, 2001: 110–25; see also the section on technical work later in this chapter).

There is increasing scope for these groups to professionalize, too. The growth and expansion of higher education institutions (as part of concerted policy drives towards a 'knowledge economy') is closely associated with the 'professionalization' of middle-status occupations. In healthcare, these are roles such as podiatrist, osteopath, radiographer and now paramedic; occupations officially known as 'allied health professions' (AHPs).

In the UK the post-1992 universities (the former polytechnics) have increasingly developed specific rather than general degree courses, offering education and training for students to gain entry into allied health professions and similar occupations. These degrees are partially academic and partially practical, with direct connections to specific occupational labour markets, and placement experience as major components. The degree of Paramedic Science is a good example. Degrees are endorsed by the AHP's professional body. This mechanism gives the professional body a degree of control over the curriculum, creating partial occupational closure while also providing a potential avenue for the profession to influence its knowledge base, deepen its skills content, broaden its scope of practice, and bid up its market value. The College of Paramedics has had considerable success with this process (Givati et al., 2018). AHPs don't enjoy the same degree of influence, status, and closure as the more traditional professions such as medicine or engineering, but they do have considerable, and growing, levels of expertise, certification, and autonomy.

The sociology of professions literature, therefore, can be thought of as a story of the expansion of 'professionalism' as a mode of organizing and operating, expanding the development of an 'expert society' ever more broadly and deeply (Becker et al., 1977: 5). Certainly this is part of it. But other literatures provide a much less celebratory picture. Critical authors characterize professions as conspiratorial and undemocratic. These elite groups—along with others—have long been highly exclusionary on the grounds of gender, class, ethnicity, religion, and sexuality (Murphy, 1988). Professions are engaged in sometimes ugly and selfish strategies of forming, seizing, and policing jurisdictions, jealously guarding 'their' realms against intruders and disruptors in the form of rival professions, ambitious new occupations, market forces, technology, regulators, managers, and laypersons (Kahl et al., 2016; Liljegren and Saks, 2017; Powell and Davies, 2012). Alongside the 'hoarding' of strategic and rewarding tasks, they also 'discard' unwanted, lower-status tasks onto support or assistant roles that tend to spring up around the fringes of core professional work (Kessler et al., 2015; Saks, 2020). Andrew Abbott's *The System of Professions* (1988) is a powerful account of jurisdictional disputes and of professional groups' largely self-interested professionalization projects. The altruistic 'trait' of a profession is contested. Critical accounts of professions note the powerful mechanisms they have used to press for and achieve positions of 'sheltered markets' (Larson, 2013/1977) and 'double-closure' of labour markets (Ackroyd, 1996; Johnson, 1972; Murphy, 1988). Professionals' special status can be abused as a shield to hide behind, to obscure scrutiny, to keep others out and to exert control over clients and competitors via quasi-monopoly powers.

Another important critical trend in the literature concerns the crisis, decline, or even dethroning of professions (Muzio et al., 2019: 25–36; Saks, 2021: 81–112). There are at least three ways in which this has happened in recent decades. The first is through professions acting in selfish ways that violate the altruistic responsibilities of public service. Examples of this include prioritizing the profit motive (O'Connor, 1998), scandals involving dishonesty and malpractice, and the attempts of professional members to cover up and deny their complicity (Currie et al., 2019). Dozens of professions have been tarnished by scandal, greed and failure. Examples include auditors which were complicit in the Enron collapse (Chaney and Phillipich, 2002); actuaries who grossly mishandled pension calculations (Collins et al., 2009); nurses, midwives, doctors, and managers who oversaw poor standards of care and silenced whistleblowers at a string of British hospitals (Currie et al. 2019; Stein, 2021); and endless corruption, safety breaches, and bribery scandals involving private corporations, government agencies, military organizations, and their legal intermediaries (Mueller, 2020). Membership of a profession clearly does not equate to a guarantee of trustworthiness and altruism.

The second challenge to professionalism appears in the form of increased managerial, regulatory and organizational control over professional workers, especially under conditions (now very common) where professionals are employed by large organizations overseen by managers and administrators (Evetts, 2011; Faulconbridge and Muzio, 2008; Noordegraaf, 2016; Saks, 2021). Managerial control and audit-based scrutiny have steadily encroached upon the levels of autonomy and self-policing that used to characterize professional behaviour. Indeed, one might say that 'management' has also undergone its own—very successful—professionalization strategy through the explosive growth of business schools and MBAs, consulting, leadership training, publishing, certification regimes, and the rise of the ideology of 'managerialism' (Khurana, 2007; Klikauer, 2013; Leicht and Fennell, 1997). Managerialism has bolstered the public legitimacy of techniques of managerial control. Management logic and language have come to dominate the daily routines of work and life, such as Key Performance Indicators, big data analytics, annual appraisals, and the language of 'visions', 'excellence', and 'agility'. Managerialism contains and restrains professionals, recasting them as just another disposable human resource to 'plug and play' into organizations that are now designed, led, and policed by a new class of professional general managers, not by the traditional professionals of the various expert disciplines. We have seen this in many industries. Lawyers in professional services firms are increasingly subjected to stressful KPI-driven evaluations (Allan et al., 2019). In automotive engineering the 'car guys' are at war with the 'bean counters' (McCann, 2016: 170–2). Managerialism has been especially powerful in public and quasi-public organizations (Exworthy and Halford, 1999) such as universities, where once highly autonomous professors are increasingly micromanaged by 'student experience managers' and 'directors of operations' (Alvesson and Spicer, 2016; Docherty, 2014; Ginsberg, 2013).

The third issue is the erosion of the value and status that society traditionally places on experts and knowledge. This is visible via three massive societal trends. The first is the inexorable rise of market-led, consumerist models of professional delivery. Clients of professionals increasingly have expectations of being placed 'at the centre' of professional service, especially when they are paying expensive fees. In the case of healthcare, patients are increasingly acting according to new discourses of 'patient-centred' or 'person-centred' care, an approach that healthcare professionals are taught to respect as something more humane, compassionate and empowering to patients than traditional healthcare models based on professional powers (Berwick, 2009; McCormack et al., 2021). The second trend is the actual or threatened replacement of human professionals by new technological systems (Kahl et al., 2016), such as artificial intelligence, machine learning, autonomous drones, platforms, and algorithms (Susskind and Susskind, 2015). The third is through the rise of 'populist' and 'post-truth' attacks on experts and their knowledge claims (Leicht, 2016; Nichols, 2017). The traditionally very powerful role granted to 'expertise' in a society is not a given. Rather, status, expertise, and knowledge claims are socially constructed. Society is no longer dominated by professions to the extent claimed by some of the more traditional literature exemplified by Friedson (1970) or Scott (1982, 2008). Expertise and knowledge bases are no longer developed exclusively by professions in their self-policed silos; rather, they emerge as contested discourses owned and governed by no one group in particular. They emerge across a range of players and actors, such as laypersons, semi-professions, technicians, managers, and self-educated patient advocates and activists (on the latter, see Eyal's work on the dramatic recent rise of autism spectrum disorder diagnoses, 2013).

Under such conditions, Evetts' (2011) notion of 'a new professionalism' is apt; 'new professions' play important social roles, but no longer enjoy (or are too young to have ever enjoyed) the degrees of autonomy and social status that professionals were once widely afforded. Traditional professions such as accounting and law are also increasing 'disrupted' by startup companies in the new fields of 'fintech' and 'legal tech' (Kronblad, 2020). We have arrived at a contradictory place. Professions and professional work play ever more important roles in an expert society, yet their status and knowledge claims are also becoming increasingly contested and vulnerable.

In this book, I use the sociology of professions literature mostly as macro-level, contextual background to the story of the paramedics. Today's paramedics in Britain certainly now 'qualify' as a profession according to traditional notions of traits and attributes, but they are very much a 'new profession' (Evetts, 2011), one located at a status position significantly below those of the traditional elite professions such as doctors, architects, and engineers. While increasingly highly trained and in many ways highly autonomous, paramedics also face powerful organizational and managerial constraints, are answerable to arrays of metrics and controls, and enjoy a relatively limited input into shaping and defining their knowledge base. Discussions of managerial and organizational control over professionals are the next theoretical component of the book that this chapter considers.

Managerial and Organizational Control

A classic image of a professional is an autonomous practitioner hired directly by a client in need of specialist services, such as a lawyer or an accountant. In reality, professionals are typically employed by large organizations, giving them a dual identity as both professionals and employees, with their latter status making them subject to various degrees of organizational control (Leicht and Fennell, 1997, 2001; Noordegraaf, 2016). Thousands of lawyers and solicitors are employed by giant law firms (Falconbridge and Muzio, 2008). High-status medical consultants see their patients within a context of sprawling, bureaucratic hospital trusts (Hyde et al., 2016). Hospitals, universities, and banks—as well as consulting, auditing, engineering, and accounting firms—are hierarchical organizations ultimately run by managers and executives and responsive to market 'discipline'. Relationships between professionals and managers are often in tension; professionals will typically resent being 'managed' by people drawn from outside the profession who are less academically and experientially qualified (Alvesson and Spicer, 2016; de Bruijn, 2010). Professionals will object when managers place the 'needs' of the market above the values of altruistic service to the public and to the professions. Manager–professional relations are complex and messy. Indeed, many professionals are also managers themselves. They embody hybrid roles, answerable to professional norms, but also answerable to the organization that employs them, and placed into an organizational hierarchy where they are also responsible for the behaviour, performance, and conduct of other subordinate workers and professionals (Currie et al., 2016).

Managerial control is far from absolute. Technical and professional workers are often difficult for managers to oversee, understand, monitor, and control. Management and leadership are often unpopular in professional fields, where managerial roles are not sought out and where senior managers are not respected. Management's potential as a profession remains unfulfilled (Khurana, 2007; Kellerman, 2018). Managerial work often enjoys precious little intrinsic interest and possesses much less social respect than that ascribed to the actual 'doing' of professional work itself (Abbott, 1981). Often looking upwards at the employment hierarchy rather than outwards at professional norms, there is always the likelihood that managers will put the interests of themselves and their organization above those of the public or the profession.

Some of the difficulties managers have in controlling professions derives from their lack of knowledge and understanding of the technical nature of professionals' work (de Bruijn, 2010), or from the fact that managers can't supervise this work in situ as it takes place in remote areas or on a mobile basis (Bacon, 2019; Orr, 1996, 2006). As a result, many occupations enjoy 'autonomy by default' (Freidson, 1970: 136). But the mere existence of 'autonomy by default' cannot protect workers from scrutiny should suspicions of poor practice arise, or if managers are sufficiently motivated to take a closer interest in observing and controlling work practices. For genuine protection from managerial intrusion, workers require the kind of formal, organized autonomy provided by governmental licensure and professional peer governance.

Public sector professions used to enjoy wide scope in this regard, but recent years have seen them being particularly strongly audited, scrutinized, and controlled (Power, 1997; Strathern, 2000). The doctrines and practices of New Public Management have affected every corner of public service provision (Exworthy and Halford, 1999; Griffith and Smith, 2014; Hood, 1991; Lapsley, 2009). The issue is one of trust. Governments want accountability for the taxes they spend on service delivery. There is always a suspicion that the private sector could do the work of public servants more effectively and more cheaply. Are governments and taxpayers being conned by the professionals and their walls of silence? Can we be confident of the value and effectiveness of their arcane, self-policed knowledge bases? Public service provision is outsourced, privatized, and marketized. Those remaining public face the imposition of a range of controls, standards, inspections, and audits designed to compel 'efficiency' and 'accountability'.

NPM has a very specific logic (Hood, 1991; Lapsley, 2009). If we want a well-educated population, it is not enough that children are legally bound to attend a set number of school hours per week. We have to account for exactly what the teachers are doing during those hours; plan every 'learning objective', measure every 'learning outcome'. Harry Wolcott's *Teachers versus Technocrats* (2003) documents how public school teachers in the USA were suddenly confronted by the forcible re-design of their practice and behaviour via flow-diagrams, checklists, and spreadsheets.

While the desire to rein in public administration and its costs is closely associated with Reaganism and Thatcherism, the lineage of NPM-style control of public professions can also be traced through the Democratic and liberal 'New Frontier', and 'Great Society' programmes of the Kennedy and Johnson era, to Al Gore's National Performance Review, or Tony Blair's New Labour and its array of targets, league tables, and penalties—so-called 'targets and terror' (Bevan and Hood, 2006; Coulson, 2009; Propper, 2008). Under NPM, professional norms are increasingly crowded out by techniques and lexicons alien to those of professionals and technicians: quality indicators, Planning-Programming-Budgeting Systems, quantitative modeling, performance targets, metrics, compliance, and audit (McCann, 2016). These modes of thinking can be difficult to reconcile with traditional occupational norms which tend to emphasize discretion, grey areas, and experiential judgment. NPM's auditors and managers aim to build independent operating systems that put them in charge of delivery and its measurement, which means installing new control mechanisms policing technique, behaviour, and even language (Chwastiak, 2001). In doing so, NPM redefines what in the organization gets measured and how, thereby forcibly resetting organizational incentives about what is valuable, what is worth doing, what gets done, and why (Bevan and Hood, 2006; Corman, 2017b; Corman and Melon, 2014). Similar dynamics occur with the development of 'Evidence-Based Practice' (Stewart, 2018), where various managerial and expert groups attempt to redefine 'what works', with the aim of more tightly designing and monitoring the practices and behaviours of public servants (for a critique, see Leah and Tomkins, 2021).

Managers, regulators, and auditors will usually try to exert ever-greater forms of control and standardization in the name of performance improvement, setting and enforcing standards, improving customer satisfaction and 'delivery', rolling out 'more effective' IT systems, mitigating risk factors, and removing operator variation and discretion. Theirs is typically a low-trust system, at odds with the professional and discretionary traditions of public service. Professionals are traditionally understood to be given a licence to practice, within which the actual 'delivery' is often more of a craft or an art rather than a science, with considerable variation and even the development of distinct individual styles of practice. Traditionally, a professional personally takes on the responsibility for effective, ethical, and appropriate service delivery, which is then rather lightly 'policed' by the professional's peers (Saks, 2021).

Audit cultures change this dynamic. A substantial interdisciplinary literature describes every aspect of the 'audit society' (Power, 1997) and 'audit culture' (Strathern, 2000) that has exploded since the late 1990s. NPM logics and indicators often appear as a notorious, sinister presence. The accountability, scrutiny, quality, and accuracy that they purport to create are widely criticized. Rather, managerialist NPM logics are often accused of distorting professional and organizational practice via 'goal displacement'. The act of demonstrating that a box somewhere has been checked, or that a target somewhere has been hit, becomes more important than the actual practices being monitored (McCann et al., 2015; Ordonez et al., 2009). Audits and performance targets are vulnerable to 'gaming' by savvy, cynical, and satisficing professionals (and managers), in a process unnervingly similar to those of Soviet industrial planning (Bevan and Hood, 2006; Coulson, 2009). This has been visible in a whole range of professions, from industrial manufacturing to teaching, social work, policy, and the military (McCann, 2016; McCann et al., 2020). Such powerful managerial influence can reformat the meaning of professionalism. NPM and other forms of managerial control create the conditions for 'new professionalism' (Evetts, 2011), whereby the very idea of 'professionalism' is recast from its traditional meaning as 'using judgement to act as an autonomous expert' to 'acting in compliance with managerial mandates and scoring highly on performance metrics'.

The full impacts of NPM and related audit cultures across and within various professions will always be contested. The most powerful professions might escape the strongest degrees of managerial control, or have the ability to resist, influence, or co-opt audit cultures to a degree that professions' established goals remain central to organizational practice (see Adams et al., (2020), Currie et al.(2016), Faulconbridge and Muzio, (2018), and Noordegraaf (2007, 2015, 2016, 2020) on the various ideal types of 'controlled', 'new', 'organizational', 'hybrid', and 'connective' professionals). But it is difficult to escape the conclusion that many professions have surrendered significant proportions of their autonomy, sometimes to a surprising extent (Alvesson and Spicer, 2016; Collins et al., 2009; Fitzgerald, 2008; Leicht, 2016; Reed, 2018).

For paramedics, their recent experience has involved *increases in both their own professional discretion and the degree of organizational control exerted over them.* Their new universities-based educational background encourages paramedics to think of themselves as autonomous clinicians. Their scope of practice has expanded considerably from its roots in first aid and transportation. But the operational pressures have also grown drastically, putting them under sometimes tight organizational and operational control. While operating 'out on the road', the demands of NPM-style audit and inspection regimes on individual paramedics are relatively light, often amounting to straightforward annual HCPC re-registration which includes the (low) possibility of a random audit of CPD profile. But the control mechanisms, standards, KPIs, and indicators applied to and by their employing NHS trust are extensive. This reality strongly shapes the ways in which paramedics are governed and deployed by frontline supervisors, middle managers, and by dispatchers in 'Control'. The sheer volume of patient demand exposes paramedics to management by stress on a daily basis.

NHS employers have also been crafty in their response to NPM demands, such as the notoriously challenging response time targets for ambulance trusts (Wankhade, 2011). Partly as a response to the 'Call Connect' initiative in the early 2000s many trusts brought in adapted road cars, known variously as Fast Response Vehicles, Rapid Response Vehicles, or what are today more likely to be called Solo Response Vehicles. (The acronym change was presumably because of the assumption that all responses should be 'rapid'.) In the USA, such a vehicle is sometimes labelled the 'fly car'.[1] The thinking was that if some crews can be split into solo responders across more vehicles, then that maximizes resource mobility and availability. The cars are substantially smaller and nimbler than ambulances, giving them an advantage in heavy traffic in urban areas. They

---

[1] 'In Bronx, 'Fly Cars' Aim to Speed Up Emergency Care', *Wall Street Journal*, 4 February 2017.

also don't transport patients so, unlike DCAs, they are not held up by long waits to offload their patients at hospital emergency departments. This means that the responder 'on the car' can be 'greened up' and made available for calls for a higher proportion of their shift time. Other things being equal, these factors should improve the overall odds of hitting response times.

But there has been controversy from the outset about the clinical effectiveness of solo response vehicles. Part of this relates to issues around measuring 'performance' mostly around speed of response, which is often not a sensible measure (Price, 2005; Wankhede, 2011). Solo vehicles are staffed by one person, have no trolley, and cannot convey patients. Many patients require transportation, meaning the SRV will have to be backed up by a DCA, potentially causing a secondary wait. In an example of the constant experimentation and churning of policy and strategy, recently many NHS ambulance trusts have steadily reduced their use of SRVs, restricting them to a small number operating in city centres only. Deployment of SRVs has also undergone change. Instead of a first response sent ahead of the DMA to 'stop the clock', it can be used more like a backup vehicle, staffed by a clinical team leader who can offer expert support and oversight, especially at complex, risky, and physically exhausting CPR or vehicle entrapment calls. Yorkshire Ambulance Service developed 'Red Arrest Teams' (RATs) which are mobilized only to CPRs. London Ambulance service employs critical care-focused Advanced Paramedic Practitioners (APPs) who operate in cars across the city. Whereas a typical paramedic might attend only six CPRs a year, one particular APP attended 270 in his first two years in the role.[2] Ambulance trust management continually experiment with changes to their deployment models in order to best aim a limited range of resources at an extremely challenging performance regime. The various impacts of these policy changes on workload and operational realities are not lost on overstretched and exhausted roadstaff.

Coupled with the unsustainable expansion of patient demand, the regime of organizational control faced by paramedics can be a hostile place for the realization of professionalism, discretion, and autonomy. In-house training is scarce and oversubscribed, the operational roster is demanding and inflexible, and managers and peers are often unavailable for support and advice (Granter et al., 2019; McCann et al., 2013). NHS trusts often

[2] https://www.londonambulance.nhs.uk/calling-us/who-will-treat-you/single-responder/ (accessed 25 June 2020).

feature a blame culture where paramedics are reluctant to discuss diffi-cult and traumatic cases with peers and superiors for fear that a debrief might reveal oversights or errors. Management is wary of patient com-plaints and bad press. Paperwork is often kept deliberately vague. The mental and physical health of paramedics and other frontline responders is often poor (Maguire, 2014; Mildenhall, 2012; Senate Education and Employment Committees, 2019).

This 'hostile environment' for professionalism is powerfully demon-strated by paramedics' high levels of self-referral to the HCPC (Van der Gaag et al., 2018). They are an outlier compared to other AHPs on this mea-sure. Paramedics and other ambulance staff labour under a blame culture and are fearful of a disciplinary culture among some parts of the manage-ment hierarchy. The high level of self-referral is also perhaps suggestive of a professional underconfidence and the practising of so-called 'defensive medicine'.

Overall, paramedics in the UK ambulance service face considerable chal-lenges emanating from the managerial and organization control of their employing ambulance trusts. Conflict and mistrust between frontline staff and management is endemic in the ambulance service, even before the days of the 1989–90 dispute. Roadstaff have always complained (both fairly and unfairly) about remote, unhelpful and 'out of touch' managers, exploita-tive working conditions, pointlessly restrictive procedures and illogical operational systems. Today's paramedics—perhaps especially those who have progressed through the university degree route—often talk of their profession outgrowing the ambulance service, making a distinction be-tween 'being a paramedic' (embodying a profession) and 'working for an ambulance trust' (being just a ground-down employee). Amid powerful managerial control and a low-trust culture, however, the paramedic role still retains significant amounts of autonomy, an issue I now explore.

Street-Level Bureaucracy

With ambulance work being mobile and operational, one of the most at-tractive conceptual sources for me to draw on for this book is the notion of 'street-level bureaucracy'. The concept was originally formulated by the public administration scholar Michael Lipsky. His famous text *Street-Level Bureaucracy: Dilemmas of the Individual in Public Service* (2010/1980) has been widely drawn on in investigations of public service organizations and occupations. It has had a particularly strong influence on research in social policy and social work (Evans, 2010; 2016; Evans and Harris, 2004; Kaufman, 2019; Maynard-Moody and Musheno, 2003; Zacka 2017).

Street-level bureaucracy (SLB) focuses on the complex interactions between client and public official. At the heart of an SLB perspective is employee discretion, experiential judgement and a deep understanding of the 'grey areas' involved in public service work. It also features a powerful appreciation of how work in public sector organizations is often hampered by shortages of resources, time, and information, with public servants necessarily engaging in workarounds, informal routines, and psychological distancing in order to manage their impossibly large and complex caseloads.

As well as being applied to teaching, counselling, and social work, SLB has also been used extensively in research on policing, another occupation where 'street level' has an especially strong resonance. An excellent text in this tradition is Maynard-Moody and Musheno's *Cops, Teachers, Counselors* (2003), a book rich in occupational stories, vignettes, and anecdotes detailing how public servants improvise and work through the complex, difficult, and sometimes absurd cases that the public presents them with. They describe how public servants 'deal with faces' (2003: 3–9). Somehow their work is simultaneously generic and specific; there are the typical categories of a case—a disabled person looking for gainful employment, a routine complaint of noise from a residential address—yet the precise details of every case, client, and context are different and each interpersonal interaction is unique. The notion of street-level bureaucracy brilliantly captures this dynamic.

Importantly, the occupations most often focused on by SLB writers, such as policing, social work, and teaching are all professions, but somewhat mid-level ones, some way below the classic professions of medicine, law, accounting, and architecture (Freidson, 1986; Larson 2013/1977). An SLB perspective demonstrates that lower-level professions are still very much characterized by the daily exercise of discretion and judgement, often coupled with the added constraint of severe resource pressures and relatively low occupational status in complex hierarchies of professions, bureaucracies, and systems (Evans, 2016; Evans and Harris, 2004; Kaufman 2019; Riccucci, 2005; van Berket, 2019; Satyamurti, 1981). They hold much of the responsibility and stress that goes with acting on personal judgement, yet few of the protections and rewards.

But SLB does not match up perfectly with the experiences of paramedics. While the 'street-level' part has clear connections to emergency services professions, there are uncertainties around the term 'bureaucracy'. This is increasingly a problematic term (Sommer Harrits, 2019). In its classical Weberian form, bureaucracy is intimately related to the ethics of office, due process, professionalism, and impartiality (du Gay, 2000; du Gay and Pederson, 2020). But in its more common use as a folk concept it is typically associated with 'red tape' and petty and frustrating paperwork. I'm not comfortable with describing paramedics (or police) as 'bureaucrats'— to my mind this really doesn't capture them at all. Vinzant and Crothers skirt this problem in their writings on frontline workers by substituting 'leadership' for 'bureaucracy'. This is in keeping with the zeitgeist, but also brings in other loaded issues. The use of the term 'leadership' rather than 'bureaucracy' or 'management' makes the often-unattractive roles of organizational control and coordination superficially more acceptable to professions and professionals (Learmonth, 2019; Learmonth and Morrell, 2019). Relabelling management as leadership can be useful in co-opting individual professionals into taking on unwanted and unpopular managerial roles. Healthcare sectors (including paramedicine) now widely use the term 'clinical leadership' (see Blaber and Harris, 2014). The rhetorical turn to 'leadership' helps to 'purify' (Collinson, 2011) management of much of its baggage as a system and interest group often seen to restrict and control professions. Whether acting as frontline operators, managers, leaders, or bureaucrats, an SLB's first instinct is usually to operate as autonomously as possible, often in ways remaining invisible to 'management' while simply 'doing what needs doing' and thereby obviating the need for management in the process. From an SLB perspective, the actual work undertaken at street level effectively becomes official policy, irrespective of the intentions of the legislators or managers who wrote the laws or procedures.

There is a tension in the literature as to whether frontline 'divergence' from or indifference towards official policy reflects SLBs' own selfish aims rather than their professional altruism (Gofen, 2014). SLBs' workarounds and improvisations are often enacted to allow themselves to handle their own heavy and tricky caseloads, and to win small battles with management, rather than as methods to best service their clients. They can falsely characterize and stereotype members of the public, strongly influencing the treatment that clients receive (Lipsky, 2010/1980; Marinetto, 2011; Seim,

2020; Zacka, 2017).[3] At the 'ground floor of government' SLBs can use excessive discretion; distorting or ignoring official policy. They can also engage in 'bureaucratic resistance' with 'bureaucracy' in its guise as something 'bad', and something quite different from its Weberian meaning of the neutral, professional officeholder (du Gay, 2000). Lipsky, while rightly emphasizing the powerful role of SLB operator discretion, often regards the improvisation, discretion, and muddling through of grey areas as problematic because managers and governments can never fully understand SLB work and can never fully control it. Such a reading of Lipsky by a frustrated politician or taxpayer would surely trigger in their minds the need for more—and better—management, audit, and control.

Other writers have taken up Lipsky in a more operator-friendly way, looking at the discretion afforded to SLBs as something to respect, something reflective of their ingenuity, professionalism, expertise, and altruism. Brehm and Gates' major empirical study *Working, Shirking and Sabotage* (1997) highlights the weak supervision of public workers yet concludes that public servants are overwhelmingly hardworking, committed, and altruistic. Indeed, Brehm and Gates suggest that the most important way in which public servants respond to government needs is through the mechanism of professionalism. Zacka (2017) describes the 'delicate moral craft' involved in the difficult decisions that a complex social world forces SLBs to make. Sometimes this involves taking a neutral, procedural stance. At other times a much more personal, compassionate, and activist interaction is the right way to go. Read in this way, SLBs are more like the powerful autonomous professionals of Scott, Perkin, and related writers (Dallyn and Marinetto, 2022). It is more in this vein that I use the notion of SLB in my discussion of paramedics. I see them as semi-autonomous practitioners; we might call them 'street-level professionals'. They (generally) place patient needs above the needs of themselves and their employers, and they rely heavily on individual decision making, hard-won experience, improvisation, and workarounds. They are often mistrustful of management and 'Control'. They despair at government targets, metrics, and standards. They want to keep it simple—focus on the patient, be human, caring, and responsive.

Their actions can put them in a vulnerable place. Public service work is often far from simple, and sometimes entails life-and-death responsibility. SLBs are often 'caught within a complex web of multiple accountabilities'

---

[3] A particularly problematic manifestation of this is 'racial profiling' by police officers on patrol (see Epp et al., 2014).

(Marinetto, 2011: 1179), and this web is a profoundly difficult space for professional autonomy and peer scrutiny, and for managerial and organizational control, especially in an era of multi-agency working where social problems range across the jurisdictions and expertise of various occupations. The work of Marinetto (2011) and Wastell et al. (2010) provide powerful insights into care failures such as the Victoria Climbié and Baby P cases, where opportunities for intervention were missed partly because of insufficient management control. On the other hand, the Francis Reports into the Mid-Staffordshire NHS care scandal described micromanagement and excessive control measures that distorted and distracted nurses and doctors from their core care duties. Matters of professional judgement can be complex to the point where it's very difficult to get the control balance right.

Evelyn Brodkin (2008, 2011) argues that, even though SLBs are increasingly controlled by NPM style audit cultures, still they exercise a wide range of discretionary behaviours. Their work can 'fall beneath the radar of management metrics' (Brodkin, 2011: i272). SLBs steer their own paths through and around NPM style procedures, just as they have always done. Josh Seim's work on California's private-sector paramedics and EMTs (2020) makes a powerful case that an SLB perspective needs to be augmented by a focus on the labour process of ambulance work within the confusing and tangled connections that exist among various public service agencies. A focus on the specific dynamics of street-level work is vital for our understanding of the roles that public services, professions and occupations actually perform in complex societies (Barley and Kunda, 2001; Seim, 2020). As with higher-level professionals, it is likely that the growth of managerial control and scrutiny means that SLBs enjoy less discretion than they have done in the recent past (Allan et al., 2019; Alvesson and Spicer, 2016; Reed, 2018). But by no means has the imposition of managerialism and NPM spelt destruction of their discretion and experiential behaviour (Evans, 2010; Evans and Harris, 2004; Ponnert and Svensson, 2016; Zacka, 2017).

SLB literature powerfully captures the indispensable and ineradicable nature of frontline operator discretion in public service. Its concerns are well connected to the realities of work for occupations and professions at the middle range, para-professional level, for public servants, and for uniformed occupations. It encourages a detailed exploration of the cultures, behaviours, and personalities of expert workers, something that is not especially prominent in the typically more abstract and conceptual sociology of professions literature. This chapter now turns to a deeper discussion of these distinct occupational features.

Service Occupations: Caring, Technical, and Uniformed

Large-scale, structural change towards a services-driven knowledge economy (Bell, 1976; Kerr, 2001/1963) is associated not only with the expansion of the professions, but also the growing importance of what might be described as 'technical' work. 'Technicians' are not easy roles to categorize and, like the semi-professions, tend to be somewhat neglected as a subject of research (Barley and Orr, 1997; Orr, 1996, 2006). Technical work 'sits at the intersection of craft and science, combining attributes of each that are normally thought to be incompatible [. . .] mental and manual skills coexist inseparably' (Barley and Orr, 1997: 12). Technicians are most certainly an occupational community (van Maanen and Barley, 1984), but are usually not understood to enjoy the status of 'professionals'.

The paramedic's roots are located somewhere in this 'technical' field. A chapter by Bonalyn Nelson (1997) in Barley and Orr's *Between Craft and Science* focuses on US EMTs. She describes the background to the US ambulance model in terms of government policies introduced to improve upon very poor outcomes for persons injured in road traffic collisions at that time. A 1966 report suggested that a 'soldier wounded in Vietnam enjoyed a better chance of survival than a motorist injured along a U.S. highway' (Nelson, 1997: 161). The EMS model was developed under the auspices of the Department of Transportation and the Department of Health, Education and Welfare. The resulting model envisaged an ambulance system based around trauma care, with a strong emphasis on rapid response and casualty extraction, with EMTs as 'physician extenders'. Responders' clinical scope of practice was tightly delineated. The Star of Life symbol was an artefact that emerged out of this process, with one of the points on the star being 'transportation to definitive care', and the language that of logistics and technical work, perhaps more in keeping with the physical and engineering locus of fire and rescue as opposed to more clinical terminology.

British paramedics similarly had their roots in a transportation and 'first aid' model, with the roles and scope of 'ambulancemen' similarly restricted. Early studies of EMS and paramedicine in the USA and UK all reflect this 'technical' focus (Hughes, 1980; Metz, 1981; Mannon, 1992;

Nelson, 1997; O'Neill, 2001; Palmer, 1983; Tangherlini, 1998, 2000). In the USA a high proportion of EMS work is carried out by fire departments in which EMS and paramedic duties are traditionally considered the least sought-after (Scott and Tracy, 2007), and the patchwork nature of the US health economy has had prolonged issues with commercial sector EMS providers. Seim's recent work documents the continuing struggles over the intense fragmentation and under-resourcing of the US EMS system (2020). Nevertheless, working in a mobile setting, technicians operate with a degree of autonomy and are a 'relatively self-contained community' (Orr, 2006: 1807), in ways not dissimilar from other technicians, or street-level bureaucrats. Technical work often involves a physical and cultural separation from management and managerial control, but professional autonomy remains fairly restricted, with resultant poor working conditions, low pay, and representation from a labour union rather than a professional association. Training is typically vocational and on-the-job. In the case of ambulance work, clinical procedures would be 'extended' to ambulance workers only after the knowledge behind them was boxed off into reductionist and procedural standard operating procedures (Nelson, 1997). Donald Metz (1981) captures this dynamic with his description of paramedics and EMTs as 'blue-collar professionals'.

Ambulance services in the USA are still engaged in a discussion about whether they are a 'trade' or a 'profession'.[4] It is difficult to see how a large proportion of first responders in the fragmented and marketized EMS systems of the USA would qualify as being employed on a 'professional' basis (Seim, 2020). And although they are mobile and semi-autonomous, ambulance workers are not beyond the scope of managerial control. New forms of technology can extend managerial control over mobile technical work, such as via satellite tracking of vehicles and the use of electronic patient records systems (Corman, 2017b). Supervisors put pressure on responders to shorten their waiting times at hospital handovers (Seim, 2020).

Although there are always limitations, the overall importance and status of the ambulance occupation or profession have changed significantly since the early 1980s (Eaton, 2019; McCann and Granter, 2019). Paramedics in all geographic settings are gaining steadily greater levels of clinical scope, autonomy, and seniority if not always better pay. Paramedics in Britain have advanced quite some way down this professional route. But anyone

---

[4] 'Is EMS a Trade or a Profession?', *Journal of Emergency Medical Services*, 28 July 2016 (accessed 25 January 2022).

working in an ambulance service today would recognize the deep cultural footprint left by the technical, uniformed, 'blue-collar' approach. There is a uniformed, 'them and us' culture, not unlike that existing in other emergency services occupations such as police and fire. This culture is often not open or progressive. There is suspicion of management, a low-trust climate, and a morbid fear of complaints, inspections, and management reprisals.

But the work does have its upsides. The uniformed culture provides its members with a powerful identity (van Maanen and Barley, 1984; Salaman, 1971), and a high degree of respect among the public (perhaps less so for police). There is also something about the culture of being involved in such a particular form of work, one that sometimes involves risk and camaraderie, that sucks a person into that subculture and makes them want to stay (Alcadipani et al., 2020; Charman, 2017; Herbert, 2006; Joseph and Alex, 1972; Moskos, 2009; Waddington, 1999). Paramedics sometimes display some of this uniformed, siege mentality; visible, for example, in the nature of the ambulance station as a sacred place beyond the range of idiotic managers and patients. This element of an immersive, totalizing occupational culture shares something in common with Coser's classic notion of 'greedy institutions' (Coser, 1974), where membership of an occupation provides a 'pattern of undivided commitment' (1974: 18).

Research on technical and uniformed occupations, however, tends not to focus on an element that is crucial to service work in general and to ambulance work in particular, namely emotional labour, and care work. These elements are absolutely central to all healthcare occupations and are particularly prominent in the sociology of nursing literature (Bolton, 2001; James, 1992), and in the broader literature on any form of customer-facing service work (Korczynski, 2003, 2007; Lopez, 2010. Healthcare work involves a mixture of emotional, physical, service, and knowledge work. For a technical, uniformed occupation that was once male-dominated (Boyle, 2002; Braedly, 2015), paramedics and other ambulance responders are mostly caring and patient-centred in nature. Paramedics are 'proud and reflexive care practitioners' (Kyed, 2020: 444). The enactment of good patient interactions depends on similar dynamics to those of the 'service triangle' of worker, manager, and customer: all three need to treat each other with respect and dignity to improve the likelihood of the service being rendered effectively (Lopez, 2010).

The changing and contested nature of the paramedic profession is reflected in a range of different types of writing emanating from those within the profession itself. The blog literature tends towards the more 'technical' side, emphasizing an anti-management position, and characterizing paramedics and other ambulance staff as downtrodden and underappreciated workers (Reynolds, 2009, 2010). Meanwhile, paramedics' own clinical literature, its trade journals, and much of the paramedic Twittersphere tends to go in the opposite direction, focusing on the ever-expanding achievements of paramedics and their profession (Eaton, 2019; Fellows, 2020).

There is continual controversy about appropriate levels of clinical judgement, responsibility, and training 'extended' to paramedics. JRCALC procedures grow in scope and depth every year. But dig a little deeper and one finds debate around the clinical protocols in use in ambulance trusts not always seen as corresponding to the 'gold standard'. Certain procedures are regarded as outdated or used unthinkingly such as the use of spinal boards and neck braces in every road traffic collision, or confusion about what to do with less common but very serious incidents, such as Sudden Unexplained Death of an Infant and neonatal emergencies (Reed et al., 2021). Confusion is also common around some aspects pertaining to mental health crises, such as the roles and responsibilities of paramedics in relation to difficult and potentially dangerous episodes such as 'excited delirium' or 'acute behavioural disturbance', or where section 136 of the Mental Health Act is invoked to detain or remove a member of the public. For individual paramedics, they often feel a heavy weight of responsibility, as their professional registration is potentially 'on the line' if healthcare episodes proceed badly. It can be difficult to stay on top of such a generalist clinical role with tremendous variety and breadth, one which can be extremely physically and mentally demanding. This leads us into the final conceptual category I use to understand paramedic work, a literature around 'edges' and 'extremes'.

Edgework and Extreme Work

Public service work of all kinds is often challenging and demanding. This is especially true for uniformed and emergency-related work, with police officers, firefighters, paramedics, soldiers, nurses, and doctors working amid accidents, disorder, trauma, crime, and the sadness, fear, and pain associated with death, danger, injury, and loss (de Rond, 2017). They are expected to be able to cope stoically with drastic and draining pressures, to stay calm, rational, and 'professional', to bracket away the human and emotional strains, and to control their own emotions and the emotions of others (Boyle 2002, 2005; Boyle and Healey, 2003). The physical, emotional, and mental challenges associated with emergency work place it squarely into the category of work in an 'extreme' context (Bechky and Okhuysen, 2011; de Rond and Lok, 2016; Graen and Graen, 2013; Hällgran et al., 2018; Klein et al., 2006). The 'extreme' nature of some aspects of ambulance work is an important part of the jigsaw of paramedic professionalism that this chapter is assembling.

There is an important and growing literature on work in 'extreme' settings. It ranges over occupations such as SWAT teams (Bechky and Okhuysen, 2011), military personnel (Fraher and Grint, 2016; Hockey, 2009; Ramthun and Matkin, 2014), warzone surgeons (de Rond and Lok, 2016), search and rescue teams (Lois, 2003), and stunt operators and film crews (Palmer, 2012). This work is not for the faint-hearted. It attracts and retains adventurous, active persons with an affinity for practical, mobile, outdoor work, often involving improvisation and teamwork. The work is unpredictable, varied, and sometimes risky and dangerous. It can be emotionally very demanding, but also carries significant rewards, such as the unique social identity it conveys, the discretion it allows, and the camaraderie offered by this unique occupational community (Moskos, 2009; van Maanen and Barley, 1984). Some texts explicitly mention extreme workers as 'trauma junkies' (Palmer, 1983), addicted to the adrenaline rush that this kind of work can generate. This trope also features in some of the blog-type literature on fire, ambulance, and pre-hospital HEMS (see, for example, Bleetman, 2012: 1–5).

The growing body of research on work in extreme contexts is an important literature. But some of it can feature a rather uncritical tone. The 'business and management' focus of some of it tends to skew this work towards finding 'lessons' from how extreme workers find ways to 'solve problems' (see for example Bechky and Okhuysen, 2011; Klein et al., 2006). This can also be seen in texts such as *Peak Performance under Pressure*, by Stephen Hearns, a trauma doctor who works for the Scottish air ambulance service. This kind of literature acknowledges the dangers and risks, but has a tendency to fall back on heroic and indomitable imagery. Extreme workers do this kind of work regularly, seek it out, and usually do find ways to adapt and overcome most of the challenges thrown at them. But there is a danger that this literature reproduces unrealistic and naïve 'heroic' tropes and narratives, as if these extreme workers experience little trouble in handling superhuman workloads and terrible emotional strain.

Voluntary risk-taking is central to Stephen Lyng's notion of 'edgework' (1990, 2005, 2014), a concept that I drew on with colleagues in prior research on ambulance services (Granter et al., 2019). Originating from sociological and anthropological writings that explore voluntary risk-taking in leisure subcultures (such as rock climbing, skydiving, or windsurfing), edgework is a form of high risk and high stakes activity that involves elevated levels of skill, practice, dedication, and concentration to master. Edgeworkers seek out this kind of activity for excitement, interest, self-realization, and to belong to a specific community. Experiencing edges and extremes can give life real meaning, and it is rare to find an occupational community where this happens during work itself. While the paramedic is best thought of as a healthcare role, clearly some parts of emergency and uniformed work could be considered edgework. These activities might include blue-light driving, resuscitation, extracting a patient from a hazardous scene, or dealing with unruly and combative patients. Certain activities involve risks not just to patients but to the responders, too. Higher acuity calls involving a threat to life are clearly more risky than other types of callout, with the daily activity of critical care paramedics or helicopter-based paramedics much more likely to involve this kind of work than standard ambulance crews. While no emergency worker would want an endless run of road traffic collisions, respiratory arrests, hangings, and falls from height, paramedics do want to be mobilized to some of these calls as they represent an appropriate use of the ambulance service and provide very high levels of meaning and involvement for responders. Part of them also enjoys and gains kudos from privately discussing these calls with colleagues on station, especially the more unusual and graphic trauma incidents (Palmer, 1983). Some of the more unusual stories enter local ambulance folklore (Tangherlini, 2000).

Some of these elements will be familiar to those associated with classical characterizations of police culture—traditionally male-dominated, a bias for action, and a closed-off 'canteen culture' (Herbert, 2006; Loftus, 2010; Waddington, 1999). Those whose work sometimes takes them into 'extreme' settings are aware of the risks, are trained to handle these risks, and they get a certain buzz out of experiencing and overcoming them. The risks are, to a degree, offset by the rewards of a job that is socially worthwhile and rarely boring. Working on a 'good job' where you've 'made a difference' can be hugely rewarding, both in terms of the actual experience itself and in what it provides to the responder in terms of a currency to trade in interesting ambulance station stories and to gain recognition as a competent emergency practitioner. This 'amazing feeling' can take an ambulance worker out of a depressive slump, at least for a certain period of time (Granter et al., 2019, 288).

While the uniform imparts a code of organizational conformity, most 'extreme' job holders and edgeworkers are independently minded and have an aversion to authority. The work is demanding and sometimes traumatic but can also be rewarding in ways that few other forms of work can. To a large extent they embrace the risks. They would much rather be out on the road than sat on station waiting for a callout, enduring never-ending meetings or wading through emails. Extreme work has both a strange connection with and an aversion to the 'hero' narrative that is often ascribed to this work by the media and laypersons (Lois, 2003).

The term 'extreme' has also been applied to jobs that do not feature physical danger and threats to life, but which nevertheless place huge pressure on their jobholders. Highly paid professional roles such as the lawyer and the business executive are sometimes characterized as 'extreme jobs' since they tend to attract high-achieving perfectionists who push themselves to extreme lengths (Hewlett and Luce, 2006). 70-hour working weeks, a rapid pace of work and terrible work-life balance are also problems for all manner of lower status and lesser paid positions. 'Extreme work' characteristics are visible, therefore, not only in 'extreme' settings such as healthcare, policing, military, or rescue occupations, but also in quite mundane kinds of work that have become highly demanding due to demand overload, resource overstretch, a collapse of workplace democracy, and toxic workplace behaviours (Granter et al., 2015). A large literature has for several decades pointed to an epidemic of work intensification across many occupations, including long hours, burnout, and widespread physical and mental illness at work (Boxall and Mackey, 2014; Green, 2004).

Ambulance work has the potential to dangerously combine two hazardous and volatile elements—edgework and extreme work. The work is inherently demanding, but parts of the organizational culture, demands, and behaviours compound the pressure and risks, making the work sometimes simply unmanageable. Hence ambulance work is a very likely scene for employee burnout, low morale, and high rates of sickness absence, especially in terms of depression, anxiety, and PTSD (Boyle and Healy, 2003; Granter et al., 2019; Henkes and Nurok, 2015; Lawn et al., 2020; Maguire et al., 2014; Mildenhall, 2012). Repeat exposure to distressing incidents can lead to 'moral injury' (Shay, 1994, 2014) another concept that has come to ambulance services from military settings. Even without traumatic scenes that are likely to forever stay in the memory of an emergency responder, the strains associated with mobile shift work in a demanding and sometimes uncivil climate are chronic. Ambulance workers are connected to a drip-feed of pressure, stress, and conflict.

Mark de Rond's work on doctors in a warzone (2017) is replete with powerful stories of 'senselessness, futility and surreality' (de Rond and Lok, 2016, 1965). Certainly that is true of war. And there is much of this in ambulance work, too. Sometimes, however, the literature on extreme work overplays the degree of trauma, and underplays the degree to which enacting care work—even in terrible situations—is 'a privilege'. An important note to add to the 'extreme contexts' literature is that much of emergency service work does not have to be 'extreme' and that large parts of it are completely routine. Weeks of shifts can go by for a paramedic without them experiencing anything particularly memorable or taxing. Many ambulance callouts don't get close to being an 'emergency' and are seen as unworthy of paramedics' time. While any callout could potentially be a life-or-death incident, the 'extreme' narrative can be overplayed, as it doesn't really capture the care aspect that is central to the identity of a paramedic as an NHS clinician. Those who make too much of the occasionally extreme nature of the work (such as boasting about callouts and seeking attention on social media) are criticized and lampooned by their colleagues. Having said this, it would be wrong to assert that the 'hero' narrative is completely absent from the work (Lois, 2013; Scott and Tracy, 2007). There is an unspoken reality that the 'action' orientation provides ambulance work with a certain excitement and even romanticism, making it substantially different from the habitus associated with, for example, nursing, and very different from many of the other clinic—or office-based AHPs, such as speech and language therapy or podiatry.

As we shall see in the chapters to come, voluntary risk-taking is often a powerful element of paramedic work. Related to this is the cultural importance of unusual callouts, and the peculiar personal penalties and rewards that come with incidents where ambulance crews are drawn into strange, hidden, sad, comical, and sometimes dangerous worlds. Storytelling and experiential learning around noteworthy and extreme incidents are important features of the ambulance service 'crew room' subculture that stubbornly refuses to die amid the development of more clinically advanced forms of professionalism.

Conclusion

This chapter has progressed through a range of discrete yet interconnected literatures on expert work. Where do paramedics fit in, especially given the significant changes that this occupation has gone through in recent decades? I argue that it makes most sense to characterize paramedicine as a 'new profession' (Evetts, 2011). Based on the traditional 'trait' approach it would be difficult to describe the 'ambulancemen' of Britain from 1970 to 1990 as members of a profession. On the other hand, one could certainly argue that today's UK paramedics possess all of the traits of a profession (intellectual knowledge base and research journals; university education; socially valuable tasks; altruistic motivation; high degree of autonomy; personal commitment to the profession; codes of ethics, see Leicht and Fennell, 2001: 26). But the reality of working life for any occupation cannot be 'read off' from items on a list. A list of key traits cannot capture an occupation's complex and contested cultural and behavioural features. Paramedic work identity is complex and multifaceted, as shown, for example, in Furness et al.'s archetypes of the paramedic as 'stretcher bearer, hero, healer, male, stoic, clinician and storyteller' (Furness et al., 2020). There also appears to be strong potential for the paramedic profession to continue to broaden its clinical focus and deepen its clinical identity. AHPs tend to model themselves on the established clinical professions. Paramedics' research journals look similar to those of the doctors and the other AHPs. They have developed other similar trappings to those of clinical professions, such as end of year award ceremonies and formal dinners. The College of Paramedics will surely soon become a Royal College, representing another major step in its development.[5]

But the profession still has some way to go in order to establish itself more fully as part of the NHS clinical landscape. Paramedics' status and roles in the wider healthcare and medical establishment often remain strangely invisible. For example, an open letter to the *BMJ* about the coronavirus pandemic on 23 June 2020 was signed by representatives of the Royal Colleges of surgeons, psychiatrists, radiologists, emergency medicine, GPs, pathologists, obstetrics, and gynaecology, and nursing, but no other professional groups.[6] Paramedics and other AHPs still aren't fully at the table.

[5] 'Royal College Status', *Paramedic Insight*, 7(3): 11.
[6] 'Covid-19: Call for a rapid forward looking review of the UK's preparedness for a second wave—an open letter to the leaders of all UK political parties', 23 June 2020 *BMJ* 2020;369;m2514 https://www.bmj.com/content/369/bmj.m2514

Status as a professional, however, is not just about formal positioning in a hierarchy of organizations and occupations. Professionals exercise their skills and judgment every day, and this is clearly shown in paramedics' use of discretionary behaviour based on education, experience, and training. The conceptual tradition of street-level bureaucracy has considerable value in helping us understand this. To some extent, the professionalism of paramedics is outgrowing the traditional parameters of the ambulance service's uniformed culture. But plenty of the latter remains.

Workload pressure is a severe challenge for paramedics, as are the operational problems: delays, complaints, workload, burnout, poor mental health, and blame cultures.[7] The work can become 'extreme', and the drive towards professional status hasn't protected ambulance staff from these pressures. It is worth remembering, however, that this situation is also largely the case for other professions. It is true that ambulance services struggle with a blame culture and workplace toxicity, but this is also a problem throughout other public occupations, such as police[8], fire[9] and the NHS, affecting all professionals from AHPs to nurses, surgeons, and consultants. There have been numerous official reports into substandard care practices and organizational failure in the NHS, such as the Francis Reports on nursing, and the Kirkup Report into unsafe practices in midwifery. Scandals around malpractice and insufficient staffing are common, and there have been numerous cases of whistleblower silencing.[10]

Professionalism is a powerful form of status and protection to have. But individual professionals enjoy this under conditions of increasing uncertainty, and often when employed in troubled organizations under strain. It is only by paying close attention to the daily realities of professional work that we can start to fully appreciate these dynamics. The following chapters lay out this story, starting with an in-depth exploration of paramedic life 'out on the road'.

---

[7] 'Seriously ill wait more than an hour for ambulance', *BBC News*, 29 January 2020.

[8] 'Whistleblower alleges toxic culture of bullying and sexism within force has left Scotland's armed police officers exhausted and demoralised', *The Sunday Post*, 22 November 2020.

[9] 'Fire services in England marred by "toxic culture"', *BBC News*, 15 June 2020. https://www.bbc.co.uk/news/uk-51127939

[10] '"I was left to fight alone for NHS whistleblower protection"', *The Guardian*, 2 October 2018.

3
Out on the Road

The Organization of Emergency Ambulance Work

'It's nice having a job where people are pleased to see you—I wouldn't want to be a tax inspector.'

Wharton (2017: 95)

This chapter takes the reader out into the complex field where paramedics work; a place where emergencies happen, where patients are cared for, where vehicles are mobilized and 'where the rubber hits the road'[1]. In keeping with the roots of the profession as a technical occupation involved in casualty extraction and transportation, ambulance workers often talk of 'road staff', and of a collective social formation of work, place and being that paramedics inhabit, something called 'out on the road'. They describe various colleagues as 'operational' and refer to vehicles and crews that have left the station as having 'gone mobile'. Ambulance services are a peculiar mixture of a clinical, bureaucratic, public service, NHS-type bearing (the language of patients, care, and Foundation Trusts) and a more industrial or military sounding world (the language of logistics, operations, response, zones, and localities). Hazardous Area Response Teams (HART) are trained to work inside the 'Hot Zone'. Crews are sent 'RTB' (return to base) at the end of their shift. Times of day are reported in 'military time': a shift will start at '0700', a crew 'arrived on scene 0310'. Radio callsigns are used to identify vehicles and crews—'Charlie 216', 'Alpha 18'. Ambulances are officially known as 'DCAs' (dual-crewed ambulances), but the crews affectionately call them 'trucks' or 'buses'. The use of the word 'crew' to describe clinical workers is a particularly vivid example of paramedics' curious hybrid identity.

---

[1] I borrow this phrase from the military lexicon, and specifically from Sir Richard Dannatt, former Chief of the General Staff of the British Army. Sir Richard is on record discussing the difference between strategy and tactics, describing the latter as the place 'where the rubber hits the road and the bullets fly', as quoted in Ledwidge (2011: 9).

The Paramedic at Work. Leo McCann, Oxford University Press.
© Leo McCann (2022). DOI: 10.1093/oso/9780198816362.003.0003

Although an employer of clinical professionals, the day-to-day language of ambulance Trusts is permeated with the words and phrases of operations and logistics. Operational management (call handlers, dispatchers, clinical help-desk staff, managers, and supervisors) work in a location known officially as the EOC—'Emergency Operations Centre'—a place often referred to by paramedics and EMTs simply as 'Control'. Geography, distance, and speed of response are important managerial factors measured by arrays of performance indicators. Organizational efforts to meet these demanding targets result in all kinds of management reforms, from changing shift patterns, updating deployment plans, commissioning different types of vehicle from DCAs to SRVs, and building and renting various base stations and standby posts. Recent years have seen widespread change to the design, location, and setup of ambulance stations; many NHS trusts have decided to close some of their smaller stations in and around urban areas and hospitals, shifting to a 'hub and spoke' model in which larger, purpose-built 'Make Ready Centres', 'Central Hubs', or 'Clinical Hubs' are located on industrial parks on the fringes of town within close range of major road junctions. Support staff at these locations are employed to manage the vehicles and their stocks, like ground crew at an airport. At the end of a shift, the paramedics' vehicle will be left in the 'red zone' of a Make Ready Centre, from where it moves through the various stages of a process to 'make it ready' it for the next crew. If there's a defect it moves to the blue zone for repairs, the black zone for fuelling, to the orange then yellow zones for external and internal cleaning, before being parked at the green zone for collection by the next crew.

Other forms of language that circulate the ambulance service are quite different from this military-industrial speak. Ambulance work is also care work, carried out by clinically trained staff with a strong bearing of helping the public, treating the ill, and acting with compassion and care. NHS paramedics have a range of backgrounds. Many have come through the IHCD route, as career ambulance workers. Others have backgrounds in other NHS occupations, such as nursing. All of these staff are intimately familiar with the humanity and tragedy of illness, injury, and death, and the vital importance of care and support for patients and their families. So amid the 'operational' talk (which largely lends itself to management), we also have the 'care' talk which tends to be more common in street-level, practitioner-focused environments. Some of this 'care talk' is technical and clinical in nature (such as the ins and outs of reading 12-lead ECGs), whereas other parts are much more human-focused, having to do with compassion and empathy. It involves a deep, very practical, and very down-to-earth understanding of how patients, bystanders, and families respond to the kinds of events and emotions that often accompany ambulance calls: pain, fear, anxiety, confusion, sadness, and loss.

The settings of ambulance work are particularly distinct from what is typical elsewhere in the NHS or in other healthcare economies. For paramedics, the work starts and ends at ambulance stations, places that are often not visible or easily accessible in the way that a hospital, or a GP's surgery usually is. Ambulance stations are often off the beaten track, found in private, almost hidden places. The station plays an interesting, but seemingly declining, role in the working lives of paramedics. Its role is in some ways not dissimilar to a fire station. In Chetkovich's brilliant study of US urban firefighters (1997), a great deal of detail is provided on local firehouses, the people who work on various watches, and the 'style' of firefighting that characterizes each particular workplace. To some extent a similar dynamic is visible in NHS ambulance services. You do hear talk of particular regions and stations that had various reputations, such as being dominated by men (or by women), or being notorious places for cliques or gossip, or for their staff being 'bolshy' and having a history of union militancy. Some stations are known to receive especially high call volumes, whereas others are supposedly 'quiet'.

A traditional ambulance station serves several functions. It is where paramedics and other ambulance staff gather at the beginning and end of their shifts. They arrive, 'book on' to a vehicle, and wait (typically not more than a few minutes) before being summoned by a bleep on the hand-held radio onto their first emergency call of the shift. Traditionally the ambulance station has often been a scene of worker control over the shop floor—management was rarely present and not really welcome on station. 'Crew rooms' are a place of banter, mutual support, moaning, joking, and resting. They are typically seen as the crews' inner sanctum, and not somewhere for 'management' to intrude into (Hyde et al., 2016: 88). There are always some large chairs for crewmembers to relax in. A TV is nearly always on. The bigger stations might have room for a pool table. There is normally a bookcase containing a few ageing copies of medical textbooks among dozens of fiction paperbacks. The sense of camaraderie on station is often strong (especially in the older and smaller stations), exhibited through informal debriefing, practical and emotional support, and gallows humour (Filstad, 2010). Staff will check up on how their colleagues are getting on, reflecting on unpleasant and difficult jobs. A few paramedics candidly described to me occasions when they have literally cried on the shoulders of a colleague on station.

But much of this is in decline. The nature and function of the station is changing. With patient demand so high, ambulance crews struggle to get back to the station at all during a shift. Once a crew has gone out on its first call, they will often then be dispatched to a series of subsequent calls and trips to hospital, meaning they can only get back to the station once per shift for a mandatory meal break. Ambulance stations have become, in the words of an ambulance trust Operational Performance Manager 'somewhere to park your car and somewhere to eat your butties' (McCann et al., 2013: 765). Smaller stations are often in a state of semi-disrepair, cluttered with strange, outdated items; operational manuals from times before the NHS trust regional mergers of 2006, and lonely and forlorn families of resuscitation training manikins. Many of the 'old school' stations have increasingly been closed and sold off as crews and vehicles are agglomerated into the newer, much larger, fringe-of-town facilities. The greatly enlarged size of these new premises can make them somewhat impersonal. Some paramedics and techs working out of the same location will rarely meet one another amid varying shift patterns. The new stations are in much better condition than most of the older locations. Some boast additional facilities such as gyms and childcare on site. The Make Ready model provides an advantage to the paramedics in that they no longer have to check their equipment and ready the vehicle before leaving the station. The development of support workers to restock and check vehicles also points to the growing value of paramedics' labour time; with paramedics on Band 6, NHS trust management will not want this time spent on menial tasks that could be completed by those in assistant roles.

But there is a downside. The reduced time to rest, to informally discuss and to 'decompress' with close colleagues is keenly felt. An unpopular development is the reorganization of work using 'standby' posts. Upon leaving the base station at the start of a shift, a crew will frequently receive a message from 'Control' to relocate to a standby position somewhere in the locality. This post will be a place where the ambulance management has done its analysis, crunched its numbers, and come up with locations on its response plan which are intended to be tactically sensible, near places of high activity or close to major road junctions for speedy arrival. Sometimes this is simply an outdoors location, such as a parking bay on a city street. It can also be a 'community response post', little more than some vehicle parking spaces and a small room that can accommodate one or two crews at a time. These can be in some of the oddest-seeming locations such as a local government office, a fire station, a police station, or even a hotel ground-floor meeting room. This repositioning can have a dislocating effect. Waiting for a call at a community response post is boring and alienating. Work and facility reorganizations, the ceaseless growth in call-out demand, and the more frequent arrival, transfer, and quitting of new staff has led to an environment where the more experienced staff bemoan the pace and degree of change. They can be nostalgic for a quieter, more grounded, perhaps more 'earthy' culture. Experienced paramedics spoke of a time when practical jokes would be played on a person transferring from a station or leaving the occupation altogether. This no longer happens. Some reflected on their experiences of moving on without anyone really noticing, the moment marked without so much as a 'sorry you're leaving' card.

Before we've even got 'out on the road' this introduction has already hinted at a range of frustrations and difficulties associated with ambulance work. Like so many other occupations in the NHS and elsewhere, there is a powerful sense of strain, conflict and deepening gloom. Questions about the value and purpose of the work are ever-present. But there is something else that is just as visible, and in many ways much more sociologically interesting. Among the complaints, the nostalgia, and the sense of general breakdown was a vital counternarrative. A very strong finding of my research was the powerful ways in which paramedics spoke glowingly about their work and identity. They often praised the variety of the work, the sense of adventure, the value and freedom of working out in the open, the sense of not knowing quite what to expect when 'booking on' at the start of a shift. While among themselves they would often joke and complain about the poor behaviour and unreasonable demands of members of the public, overwhelmingly they spoke about serving the public and treating patients as 'a privilege'. This idea held great meaning for them. Sometimes it would emerge in discussions of genuinely life-saving emergency calls. This sense of 'privilege' would also be present even when—perhaps especially when— a patient had died. At times when the outcome of an emergency call is bad, the fact that paramedics were present, did everything they could, and tried to comfort the bystanders and family members was 'a privilege'. Among all the stress, strain, and conflict, this element is of profound importance to ambulance work and, as we shall see, something that features regularly in the analysis to come.

But for now, the present chapter will focus mostly on the more mundane elements of paramedics' daily work duties. It explores the daily routines of ambulance work, focused through my understanding of the paramedics' experiences based on my own observations, and on points and arguments made in detail by paramedics in the individual interviews. It will firstly explore the troubling discussions over scheduling and rostering that, given the intense overstretch throughout the system, are contentious matters in every NHS environment. The chapter then describes what happens when paramedics are mobilized to an emergency call, including how they drive to the call (while often being diverted from one 'job' to another). Implicit in this description is the paramedics' discussions of how well (or more likely how poorly) their work is organized by their employing NHS Trust and by 'management' in general. As we shall see, in some sense ambulance trust management employs robust forms of control over the various 'resources' it has 'on the road'. But there are many powerful limits, blind-spots and restrictions as to how this 'control' actually operates. As professionals and workers in many occupations will attest, a crucial ingredient in making their work engaging and survivable is that work activities should be, as far as possible, meaningful—that at least some of the hours and years spent working in an organization should (wherever possible) involve discretion and respect for operators' skills and abilities and that all this effort should ideally achieve something tangible and beneficial. What professionals of all stripes often resent most of all is their time being wasted on 'busy' work that distracts from their 'real work', whether this is teaching, social work, coding, accounting, policing, architecture, or machining. One of the things I found most fascinating in my years spent researching the paramedics, was that the meaningfulness of their work varied intensely on this scale. In addition to the work being worthwhile, engaging, and vital for the public, there are also strong themes of the work being exhausting, pointless, and dreary. Central to this was the frustrating sense of being moved around 'from pillar to post' while achieving very little of note for patients and the public.

The specific areas that this chapter explores are as follows: (1) work time and the organization of shifts; (2) driving and handling information; (3) working with other crews; and (4) treating patients and relating to the general public. The chapter provides an empirically informed descriptive overview of the NHS paramedics' duties so that the reader has a basic foundation of the nature and organization of their work. Later chapters place more ethnographic description and analysis onto this foundation; narratives that go into much more depth into conceptual areas of interest: namely occupational culture, stress and coping mechanisms, and the contested meanings of professionalism.

What this first empirical chapter aims to demonstrate above all is the sheer broadness of the emergency ambulance working experience. Paramedic emergency response is, by necessity, extraordinarily wide and varied. Certain calls require specific application of clinical and practical knowledge, sometimes involving considerable degrees of clinical risk. Others require very basic interaction with patients, involving nothing more advanced than taking basic observations, and providing the kind of human-centred care, support, and advice that a friend or a family member could have provided if they were available. Very often, ambulance work involves frontline discretion, improvisation, and experiential decision-making. It also requires very advanced social skills, compassion, and patience. Fundamentally it is about care for people in need. Paramedics must possess a practical, hands-on ethic, and a real commitment to working with the public. With time, they also become skilled political workers, finding ways to navigate their patients through the organizational and occupational corridors of the NHS as they try to provide them with access to the most appropriate forms of care available. They are also expert practical problem solvers, having to daily find ways to drive emergency vehicles at speed to arrive safely through difficult conditions, figuring out how to move people around obstacles, finding ways to safely take charge of a public scene. Certain conditions require the need for an authority figure, to advise or even to command others, not unlike a police officer, that classic example of the street-level bureaucrat. Others need simply a sympathetic, caring person (or two) to help a patient in some form of medical, psychological, or social distress. As with other frontline workers across the public sector (Evans and Harris, 2004; Maynard-Moody and Musheno, 2003), street-level discretion is central to the life of the paramedic. Instances and descriptions of autonomous decision-making and action appear regularly in the interview and observation material. Discretion is also central to professionalism, and it will be explored in more depth in Chapter 6 as the book heads towards its conclusion. But we have to start at the beginning. Our empirical explorations begin with an account of the places where paramedic work usually starts and finishes on any given shift—the ambulance station.

'Booking On': Work Time and the Organization of Shifts

Shift patterns differ across Trusts and regions, but twelve-hour shifts on the road are the 'bread and butter' of ambulance life. Paramedics and those in other operational roles such as EMT or ECSW would typically take on around three 12-hour shifts per week. These are often compressed into a series of shifts, such as three daily 7 a.m. to 7 p.m. shifts in a row. Crews talk of being 'on a 3 x 12' or 'a run of nights'. Discussions of long shifts, anti-social hours, and exhaustion were a common feature of my interviews:

> Gareth:[2] The shift patterns are horrendous [. . .] Physically it's tough. I'm getting older, my joints are fairly stiff. This is having a detrimental effect on my life. The tiredness is hard.

It was not only the more experienced paramedics that complained of the pressures that shift work places on the body and mind. David, a degree-trained paramedic with around six years of experience, claimed that 'it really does mess with your body clock'. He went on to describe the 'up and down' nature of tiredness, expressing concern that the exhaustion associated with shift work could potentially be dangerous for patients and staff:

> David: I've been so tired recently. I was driving back on the motorway after finishing a shift, and I was nearly falling asleep at the wheel! I had gone one hour over the end of the shift into unpaid overtime and I was really knackered. You wonder if it starts to affect your practice. I was so tired on that shift, I was eating sweets, drinking coffee. But also being a bit sleepy helps you stay calm. A lot of the time you can cruise; the jobs are often very routine, like a hot baby, or abdo pain, specific jobs are really easy and there's nothing to do except help the parent or the patient relax. But you can relax too much—you switch off. Then you go to something more serious and you get this kick of adrenaline, it helps you cope.

Clinical research has demonstrated that shift work disrupts a person's daily physiological cycles or 'circadian rhythms', leading to a range of unwanted and possibly hazardous outcomes, from disturbed sleep and reduced

---

[2] All names of people, places and organizations are pseudonyms. For details on the research process involved with the writing of this book, please see the Appendix.

human performance, to increased likelihood of physical and emotional illness, including serious illness such as cardiovascular disease (Kuhn, 2001). In addition to the challenges of shift work, the mobile nature of pre-hospital operations exposes operational ambulance workers to enforced overtime, something they really resent and always hope to avoid (Seim, 2020). This often occurs when a crew is dispatched to a case close to their shift finish time. A care episode cannot be abandoned once committed to, and the demands of a call received near the end of the shift can push a crew well over an hour or more past their shift-end time. The dislocation caused by anti-social work hours is also aggravated by the varied location of ambulance work, which is unpredictable and fluid. The ambulance station provides a clear 'workplace' for ambulance crews, especially when they arrive to 'book on' to a shift. But, once they've left to answer their first call, the pattern and location of the day or night's work can be extremely varied and impossible to predict. Stations are one of very few ambulance work locations that are set in one physical space.

Below I describe the scene at an archetypal 'old school' station before going out on an observation shift. The station was a one-storey building, consisting of a large garage abutting four small inner rooms: a cramped office, a larger mess room with kitchen, TV, easy chairs, and a bookcase, and an adjacent lockers and toilet area. The facility was located on a ring road in a medium-sized town. It was unobtrusive. A pedestrian could easily walk by and completely miss it. I found the start of this particular observation shift to be unusually quiet. Times on shift without a call are rare. They were invaluable for me to take stock of what had been going on, to talk to other crewmembers and to catch up on fieldnotes. For the crews, moments of inaction were helpful for resting and conversing with colleagues. But, as my fieldnotes indicate below, there was clearly an issue of boredom, usually signified by a shrug of the shoulders, a joke; a sense of bewilderment about what is happening. Why haven't Control assigned any jobs? There must be some mistake.

> I arrive at the station and meet a couple of staff, including Sally, a tall, imposing ECSW, whom I'd come across before–a fan of trashy horror novels. She is sat in the little office area reading a Tom Clancy. Chris, the CTL, sits in the rather dark and dingy rest area talking to me about the geography of the area and the ever-rising demand.
> There is some sort of bleep and activity on the radio so we head to the vehicle. But as the crew look at the MDT there is nothing there. 'I must be hearing things', mutters Chris. Sally seems to find this hilarious. She laughs at Chris,

setting the tone for the rest of the day—she has him fair-pegged as a chaotic, scatter-brained boss who is impractical compared to herself as the less senior but more practical technician.

We head back to the mess area and Sally sits back down with her book. 'I must say I'm enjoying this observational shift!' she calls out jokingly from the office area. It is odd how little there is for us to do. It's been almost an hour without a job. 'We call it "the curse of the observer"', says Chris.

Day-to-day chatter and gossip about shift patterns, crewing, rotas, holidays, and tiredness is pretty much constant. The following fieldnote illuminates how paramedics interact with one another as one shift comes to an end and another begins, with work time and daily organization of the work a central theme. I was about to start an observation on a 7am–7pm shift with a Critical Care Paramedic crew. I have just met the CCP named Becca and we are heading out of the station to the vehicle to prepare for the shift to start. Her partner hasn't yet appeared, and we were also waiting for the other CCP ambulance—or 'truck'—to arrive back at station at the end of their shift so they can hand over the advanced airway bags and the 'Lucas' or 'autopulse' machine—an ugly, unnerving piece of equipment that provides mechanical CPR to patients in cardiac arrest. Becca is pottering around in the back of the ambulance, breezily chatting to me about my research project.

She points at the trolley—'looks like someone's been sleeping on this!'—and tidies it up a bit. Another ambulance parks up alongside us. It's the other CCP crew coming off the night shift. I only really see one member of that crew: a tall, young-looking man with glasses and dark hair. His demeanour is agitated. He looks bothered and haggard, keen to get away. But before he goes he talks extensively with Becca about shift patterns, holiday leave, etc. He seems exhausted.

The rear door of the ambulance opens and another man in green jumps in. I guess this must be Becca's partner for the shift (Pete), so I introduce myself. He says he's being doing the job three-and-a-half years. He's coming back into it after some time doing something else. 'It's been tough to get back into shift patterns, working nights, you know? It's taken me a while to adjust back to it.' Becca says she needs to head back inside the station to sign off on some paperwork, so we wander back inside. As the other CCP heads away (presumably to his own car to drive home to bed), I hear him continue to grumble about shift patterns. . ... 'When you're off, that's when I've got my fucking bollocks of a shit week.'

Constant geographical movement (sometimes over long distances) can also be disorienting. Later on that same shift, Pete reflected on this with me in a lighthearted way:

> We have a patient in the back being attended to by Becca, so Pete and I are talking in the cab as we make our way to the hospital. It is slow going. Traffic is starting to build up by now.
>
> Pete: 'I thought I was being clever coming this way, but I've got us stuck in traffic, I forgot it's rush hour now. You get like this with shift work, you forget what time it is—like at night-time or weekends. Everything's shut! I remember once I'd said to someone 'I'll call you when I get a minute at work today.' But I was on night shift and I nearly rang my mate at 2am and had to stop myself - what am I doing?! It's like, 'I need to go to Boots'—'What? It's shut, mate, its 4 in the morning!"

But, as ever with ambulance work, the complaints, absurdities, and grievances were wrapped into a broader narrative about enjoyment, professional autonomy, and of doing useful work for the public, as noted here by paramedic Rachel.

> Rachel: I like to have continual callouts to pass the time. I like the variety, it can be fun, who you are crewed with, what they are like, what makes them tick, how you interact. [. . .] [I]t's the best job I've ever done. It's never boring.

'The best job I've ever done', 'the best job in the world', and doing a job that 'makes a difference' were extremely common tropes, resurfacing again and again in my interviews and observations. The enjoyment comes from several sources. Part of it is the professional pride associated with serving the public. It also comes from the variety of the work and the fact that crews work outside rather than being cooped up in an office. This varied work, out among the public, also creates an element in which ambulance work—like other uniformed occupations such as policing and military—can be bizarre, sometimes involving amusing, anarchic, and unpredictable events. Such notions appear regularly in the 'blog' type literature written by paramedics or other ambulance staff themselves (see Reynolds, 2009; Wharton, 2017). Very often I heard strange and unusual stories detailing the infinite possibilities for human beings to involve themselves in dangerous, bizarre, humiliating and crazy situations. Timothy Tangherlini's anthropological work on US paramedics' storytelling focuses almost wholly on these colourful, often 'extreme' stories (Tangherlini, 1998, 2000; see also Granter et al., 2015). Often these stories feature 'black' or 'gallows' humour as a method of coping (Charman, 2013, 2017). Some of the stories I heard will be shared with the reader (where appropriate) in later chapters. I discussed this element with Emma, a highly experienced paramedic, in one of several long interviews I had with her.

> Leo: It can be fun being out on the road, not knowing what you're going to get next. It is varied. I mean, personally, I was often fearful the night before and I found it exhausting, but it was also exciting.
>
> Emma: Yeah, 'exciting' is maybe not the right word. I wouldn't call it exciting. But, yeah, you're up for it. You know: 'let's see what's coming.' The novelty of the blue light driving does wear off. At first you love it, passing red lights, flashing speed cameras and just laughing about it knowing you're not gonna get fined! But it does wear off, it just becomes completely routine.

Emma's 'completely routine' and David's 'a lot of the time you can cruise' are very accurate statements about much of the work—at least for paramedics who are experienced and physically and mentally well. While ambulance work is certainly tiring and physically disruptive, large sections of shifts, an entire shift, or even a week of shifts can go by featuring only mundane activity. In addition, after building up the requisite experience, almost anything can start to feel routine. Travelling to 'jobs' on lights and sirens, overtaking lines of traffic and proceeding through red lights becomes totally unremarkable when performed 10 or more times in 12 hours, week after week. The siren noise gets to be annoying. Ordinary road users or pedestrians see a speeding yellow ambulance and hear its wailing cry as something exciting, perhaps even romantic. They might assume that the heroic occupants are rushing against time to save a life. They might wonder 'how do they do this job?' But the paramedics and techs in the vehicle are not going to be thinking this way. For them, blue-light driving is simply an everyday form of work activity, something I explore in more depth in the next section.

'Going Mobile': Receiving and Responding to Calls

A large proportion of an ambulance shift involves driving or riding in vehicles. The purpose and nature of ambulance driving has several forms.

Obviously it includes driving on lights and sirens en route to emergency scenes, a form of driving with unique challenges and features. But it also encompasses transporting a patient to hospital (it is relatively rare for a patient's condition be serious enough to warrant using lights and sirens on the run to hospital), and simply the very mundane acts of driving the vehicle 'RTB' (return to base), or relocating it to a standby post, which is done in exactly the same way as driving any ordinary road vehicle.

Very little academic work exists on driving as an element of EMS work, with the notable exception of Corman (2017a). Technical training manuals emphasise the need for caution, restraint, professional standards and, in an interesting example of managerialism, of treating the entire gamut of emergency response—the driver, the vehicle, the environment, the call—as 'a robust driving response system' (AACE, 2018: 29). Similar police service documents talk of 'roadcraft' (The Police Foundation, 2013). These texts have a didactic, technical, legalistic, and 'operations' style, with passages, for example, warning the emergency driver against 'emotion-focused coping' or 'confrontive coping'. Instead the correct approach is 'reappraisal coping: dealing with driver stress by re-appraising one's emotional and cognitive reactions, which tends to have a more positive influence on driver behaviour' (AACE, 2018: 30–1). The reader learns of certain skills and procedures for general and specific matters, but is also reminded of the need for professional discretion, individual responsibility, and common sense: 'As there is no predetermined procedure for every conceivable type of situation that develops, drivers must continually perform a dynamic risk assessment on the changing environment and conditions' (AACE, 2018: 29).

Although done so often by operational staff as to become totally mundane and unexciting, emergency driving on lights and sirens is an activity requiring technical training, intense concentration, and no little skill. Driving at speed and legally through red lights and the wrong sides of junctions involves increased risk to all road users and added responsibility for the emergency driver. Several of my interviewees mentioned their unease and frustration about this element in relation to the issue of medically unnecessary callouts. Many 999 calls are made for relatively minor ailments that are clearly not 'genuine' emergencies. Yet if any call translates into an urgent dispatch for a crew, then that crew has no choice but to drive on lights and sirens as if it is a 'genuine' emergency. Although some paramedics and other ambulance personnel will exercise a degree of informal discretion over this, others will follow management procedure to the letter and use blue lights for every call, no matter how seemingly mundane. As Karen, a paramedic of ten years, explains:

Karen: Some crews do take risks, really. They will interpret the call as less serious than it is and maybe take it a bit too easily. I make sure I don't do this and
respond to every call the same way. It's not worth the risk of something going
wrong and then having to explain yourself.

Medically unnecessary calls are a massive problem, notorious throughout
the EMS world. Sociological research on paramedics in North America
portrays responders bemoaning 'shit calls', or the comparison between 'le-
git' and 'bullshit' callouts (Corman, 2017b; Metz, 1981; Palmer, 1983; Seim,
2020). Conflicts about 'pointless', 'rubbish', 'dross' or 'nothing' calls, and
how 'Control' (mis)manages them are similarly very profound in England's
ambulance service, with this issue covered in depth in Chapters 4 and 6. But
for now, it is worth point out that the issue of 'unnecessary' calls also connects with driving, in that calls interpreted as pointless expose ambulance
crews to a needlessly expanded volume of work demand and therefore expanded legal and physical risks. While sat at an ambulance station between
calls, I listened to a young paramedic named Josh reflect with me and other
crewmates on this issue. As was very common in talk of this kind, the discourse is framed within an explicit critique of the employer. In Josh's very
Lipskian view, 'management' and 'management policy' does not understand the 'reality' of the 'demands' of the road, meaning that the street-level
responder has to use his or her discretion to work around an unclear or
contradictory system:

> There's a difference in the law and the management policy, you know that,
> don't you? I was always taught to think 'can you explain in court why you used
> blue lights on a job if you've hit someone [been involved in a road traffic
> collisinon]?' But the policy pretty much tells us to use blue lights on every job.
> What if my explanation in court is that I just used blues for every job, regardless
> of how serious it seems or doesn't seem? It doesn't help when a cut thumb
> comes to you as 'penetrating thumb wound.' I didn't go on lights to it. Forget it,
> I'm not risking ploughing into a pedestrian for that. No way.

While ethnographic enquiry is often claimed to be vital in showing us 'how
things work' in organizations (Watson, 2011), I found that my fieldwork
was never able to fully explain this particular issue of informal decision-
making around perceptions of 'genuine' urgency and patient need. While
Josh's critical and off-the-cuff account is here counterposed with the more
conservative and rule-following Karen, it was impossible to draw any clear
inference as to how exactly paramedics and techs decided how urgently

to drive to specific calls, beyond pointing out that this differed according to time, crew configuration and, it seems, employee mood. According to rule book and training, they almost certainly should not be exercising any discretion over how they interpreted the urgency or otherwise of a call. But, in practice, some of them did seem to do exactly this, or at least allude to doing it. In any case, the rule books and training manuals were themselves confused. They simultaneously claim that all calls in a certain category must be driven to in emergency mode on blue lights and sirens, while also suggesting that responders must always judge for themselves the conditions and situations. This was one of several grey areas around discretion, interpretation, comportment, and style of practice that differed across the various personalities I came across, something clearly in line with the expectations of an SLB perspective (Lipsky, 2010/1980; Vinzant and Cruthers, 1998; Zacka, 2017).

In their driving duties, paramedics also interact with members of the public other than patients in the form of 'other road users'. As one might expect and hope, I tended to find that 'other road users' were usually diligent and cooperative when emergency vehicles were trying to make their way past traffic while running on lights and sirens. But this wasn't always the case. As the following fieldnote about a blue light run in heavy traffic explores, the lack of consideration shown by other drivers is a common bone of contention for ambulance workers. I'm sat in the back of the ambulance, craning my neck around to look through the portal into the cab:

> Eventually we make it out of the town. We're soon in a rural area and pick up a lot of speed. At one point we're doing nearly 80 miles per hour and Sally catches the rear left wheel on a kerb - that corner of the ambulance really bounces and, although I'm sat in my seat, my upper body is thrown sideways into the cabinet next to me. I've got a seatbelt on, but there is so much sideways movement that I often have to 'hang on' a bit.
> At one point Sally slows and aims to make a right turn into a country lane. However, not one, but two oncoming vehicles decide not to give way to the ambulance on lights and sirens and instead take left turns into this same lane ahead of us! I was pretty surprised by the first one, but the second just took the piss completely. 'Ohhh, seriously!?' shouts Sally but she just laughs it off.

At a quieter moment in the day, I asked Sally about the standards of other people's driving:

I've once been overtaken and the driver gave me the finger—unbelievable how
some people drive. BMW drivers are definitely the worst—they drive
aggressively and never indicate. Mercedes - they're slightly better but often
bad.

Poor standards of driving around emergency vehicles is an annoyance. It can also be dangerous. Reported road traffic collisions involving ambulances in the UK happen as often as 25 times per day; fortunately the majority are ambulances striking parked vehicles, whereas potentially dangerous collisions at junctions make up only about 5% of the annual total (AACE, 2018: 6). Infamously, irate members of the public have been known to leave complaint messages on the windscreens of parked ambulances which have blocked roads or driveways, something that news media have picked up on extensively as a sign of an individualized society with no moral compass.[3]

On occasion, an ambulance will have to drive extremely fast if the crew is some distance from a life-threatening emergency. As an observer, I experienced this on a few occasions, such as in the following fieldnote when I accompanied a CCP crew to a CPR call in a nearby town that required a motorway drive to get there. Throughout the ambulance journey I was struck by the intense concentration yet relaxed state of the paramedic driving—it was Becca on this occasion.

The MDT gets more information: 'Male, DIB, muscle problems, turning a funny
colour.' Pete: 'That's the standard of medical terminology we have in the
ambulance service. Love it.' His comment lightens the mood a bit, and all of us
laugh as the vehicle heads away from town and towards the motorway. With
lights on and sirens blaring we are getting ever closer to a Nissan Micra that is
pootling along at about 20, 25mph. We come up to about fifteen yards of it,
blue lights rebounding off its paintwork. The driver still doesn't seem to have
noticed us. 'Jesus wept', says Pete: 'It will be a little old lady who can't see over
the steering wheel. . .' Eventually the driver notices and pulls to one side. We
are quickly out of the town and on to major 'A' roads and ring roads. We are
really picking up speed. Cars are all moving out of our way. I notice the way
that our lights reflect strongly back off the white reflector bands around the
outside of the traffic lights—they go a shimmery deep purple. It gets a bit tricky
at roundabouts but Becca finds a way through—sometimes it is pretty tight

---

[3] 'Abusive note left on ambulance in Southend', *BBC News*, 15 November 2019.

when everything slows and bunches up. But mostly the crew is making very rapid progress. The miles to destination are flying down on the sat nav. We get to a junction and Becca slows right down for it—in doing so she somehow finds the time to explain to me that there is a very nasty blind opening just ahead and that sometimes traffic just emerges straight out of it; drivers can't see traffic to their left and should obviously give way, but sometimes they don't! I was very impressed by this. The experience of emergency driving, years on the road, combined with local knowledge of unusual or dangerous road layouts—metaphors of an expert sportsperson or video gamer come to mind. Performing at this high level is a mixture of instinct and ability but also the product of hard-won experience: planning ahead, learning from past events, concentrating hard. It's all functioning at such a rarefied level. In some sense it really was operating the vehicle 'as a robust driving response system' with Becca conducting 'a dynamic risk assessment'.

We make rapid progress amid traffic that periodically bunches up at junctions and roundabouts. The ambulance slows to a crawl to negotiate around cars that are stuck, unable or unsure of how to move out of our way. Becca cycles through the repertoire of siren tones; the vehicle barks out different songs from shrieking to honking, almost to jog uncertain road users into compliance. She eventually reaches for the extra-loud horn, which is activated by a red button next to the steering wheel. 'BAH—BAAAAAH': it's immensely loud and has a satisfyingly grumpy and aggressive tenor, like the sound of fire trucks in U.S. action movies.

Major roundabouts just before the motorway are a little hair-raising. The ambulance leans very noticeably when steering at speed. A funny sound comes from the dashboard, kind of a growling, beeping, complaining kind of noise. Apparently it's the traction control warning. It even does this once or twice under braking which gives an impression of how fast we're going at certain points. It is a weird noise, almost reminiscent of the old grumbly sound of dialup internet. It's a bit of a feeble warning or complaint. Like several other electronic and radio sounds in the cab the crew seems to have basically ignored it.

After a couple more of these large roundabouts we are soon bombing down the entrance slipway to the motorway. Here we go. The vista opens up of the three-lane motorway. Plenty of traffic: cars, vans, lorries. Becca takes us across the lanes so that we are in the third, inside one. Everyone ahead of us is sensible, and just moves across to the middle lane. Our speed picks up, 80, 90, 100. The big, heavy ambulance's diesel engine is really roaring now, doing everything it can to keep the speed up. It's not going to go any faster, somehow

I can sense that the vehicle can do no more. We notice up ahead that there is one car that we're not gaining on! We were behind him for ages and we weren't getting any closer. Becca laughs and comments on this, now also mentioning herself that the ambo won't go any faster. I'm very impressed by her driving abilities and her relaxed but concentrating demeanour. Bizarre as it sounds I think for a moment about O'Neil's research study of ambulance work in that chapter of *Inside Organizations* when he mentions racing down the motorway and wondering to himself why he's put himself at risk like this. This is in the back of my mind, for sure. But I am also completely calm and don't feel fearful at all. I quite enjoy it, to be honest. Like I have with other paramedics, I have complete faith in her driving ability, and I can see the impact the lights and sirens have on other drivers. People are keeping well out of our way. With the sirens on for so long they start to sound different to me. Not in that sense of a Doppler effect, but more a product of that odd process where you've been exposed to the same repetitive rhythm for so long that your mind starts to tune into different timings within that rhythm; the tone starts to sound to me like a growl, there's a kind of whirling and warping to it. Motorway signs mounted on the bridges are bathed in a strobing deep blue, almost an ultraviolet, that comes from our lights. It is a peculiar sight; motorway signs are extremely mundane and everyday, but here I'm seeing them – literally – in a different light. It's like I've been transported to another place.

As the reader can probably tell from that extensive and colourful fieldnote, the experience of genuinely flat-out blue-light driving was a rare one for me. Reading this text again, years after the event, the experience comes over as exhilarating, indicating awe, nervous energy, and an adrenaline rush that I can remember experiencing. It reveals the kind of vividness that is sometimes the product of the naivety of the academic ethnographer making his or her diligent fieldnotes of an exotic, intoxicating culture. Perhaps a student paramedic might feel this way on their first experience of observing an emergency of this kind. For crews, however, my reading was that they probably enjoyed this kind of intense driving but experienced it within an employment context that rendered it 'everyday' and therefore not worthy of much reflection. But this kind of driving—the skill it demands, the responsibility it entails—is most certainly challenging. Despite it becoming 'everyday' (especially for CCP crews) it strikes me as neither sensible nor healthy to causally dismiss or downplay its intensity. Blue-light driving is hard, demanding, and potentially very hazardous work. Even when not responding to the most serious, potentially life-threatening calls of the kind described above, blue-light driving usually requires a strong degree of multi-tasking en route (Corman, 2017a) and places a significant cognitive load on the driver. A non-driving crewmember can be very helpful in reading and acknowledging updates that come through on the MDT, and possibly discussing details of the call on the radio or phone if necessary. When working solo the paramedic has to do all of this alone, making the work all the more demanding.

In a classic article (Hughes, 1980), medical sociologist David Hughes describes the 'ambulance journey as an information generating process' within the context of a patient being transported to an emergency department in an ambulance. That paper is very much a product of its time, a period when there were barely any paramedics in the UK, only 'ambulancemen' whose clinical input was limited to finding out information which might be of use to the definitive medical carers at the hospital—the much higher-qualified doctors. This notion of the ambulance ride as a clinical information generator is interesting, but such are the degrees of change, intensification, and expansion of ambulance services roles, that the 'information generating journey' often occurs en route to a new scene, before the paramedic has even met the patient. With the dispatchers or clinical advisors working at the EOC sometimes in continual contact with a patient or bystander, further patient information is often relayed via the MDT to the ambulance crew as a unit drives towards the scene. Often the additional information leads to a downscaling of the call, leading to the responding crew being diverted to another call that is possibly more serious. Thus, Hughes' 'information gathering' occurs even when there is no ambulance journey, at least, not for a patient. Some shifts melt into a frustrating series of never-completed journeys for a crew; a digitally mediated, blue-light, wild goose chase. This tends to be particularly common for paramedics working on solo-crewed vehicles. The observation fieldnote below illuminates an experience that a solo-working responder can grapple with several times per shift. In this case, after all this intense driving effort, there is not even a patient at the end of it.

> We get a call after having been stationary for about ten-fifteen minutes talking in the vehicle while sat on the ambulance station driveway. It's 1202. We're heading to a call described as a 'male suffering from abdominal pains'. We set off on blue lights and merge onto the ring road heading out of town. The traffic is heavy but we make relatively rapid progress and aren't held up much by traffic blockages. We seem to be progressing on blue lights for a very long time;

the job is about 15 miles away and it takes about 20 minutes to get there. We get further information on the radio, an update on how the call is progressing, suggesting a degree of urgency. This would have been classified by the EOC as 'Red 2', potentially life-threatening with an 8-minute response time: The radio comes alive: '[TRUST], we have patient with abdominal pain, possible AAA, we have RV running from Bonneville [that's us] and DMA running from Housestanton.'

The MDT keeps making a 'binging' noise and paramedic Emily asks me to hit the 'Ack' button to acknowledge. Fragmentary bits of further information filter through the system. Quite how a solo operator is supposed to keep up to date with, digest, and acknowledge all this while driving on blues is beyond me. Emily is concentrating hard on her driving, much of it on busy A roads or through town centres clogged with traffic and pedestrians.

'Now, this I do find stressful', she admits as we close in on a busy roundabout, confronted by traffic waiting for us and (mostly) giving way. All the while the siren is wailing, the Control guy on the radio is monologuing and the MDT keeps making 'binging' noises asking us to 'Acknowledge'.

Emily: 'Sometimes you just have to wait until the road is clearer before replying on the radio. You can't do three things at once!'

'Bi-dung, bi-dung, bi-bung' goes the MDT. I hit 'Ack' again, while trying to take in the amount of things happening that Emily has to handle: getting closer to the job, making sure she takes the right turnoff, making sure traffic is indeed giving way to allow the vehicle through. We wave to other ambulance crews that we see from time to time.

With blue lights still on after around 12 minutes we finally seem to be getting very near to the job. We're in a small town progressing through tightly laid-out residential streets. Parked cars on both sides of the road provide limited room to manoeuvre. Emily says that she has sometimes been on really long drives on blue lights, occasionally for as long as 40 minutes.

40 minutes. . .? I don't ask her anything more here, but I assume this must be when she's travelling 20 minutes to one job, then gets cancelled down and sent to another one straight away.

Similar details emerged in an interview with Ben, a senior manager at a Trust who wore the green uniform to work and booked on for operational shifts on the RV at least once a month. He described the issue of paramedics spending ever-longer times on scene, and described one of the 'jobs' he had been to recently on a shift where this was 'the only job with a patient'. I asked for clarification:

Leo: What did you mean 'only job with a patient'?

Ben: Oh, it was the usual thing of the jobs, just racing around but never getting to a job. The job comes in, the dispatch goes out almost before he's finished looking at it, the call is sent to the vehicle, the crew heads off towards the job on blue lights, meanwhile the call handler takes some further information. By this time it becomes clear the call is less urgent, then its downscaled further to 'hear and treat'. The vehicle is stood down, then is immediately sent to the next job. You get close to that job, then another comes in, you're diverted, on it goes. . .

The problem of 'racing around but never getting to a job' is endemic to emergency ambulance work and speaks to a deep problem of frustration. It is connected to complaints about 'Control' and 'management' who, for street-level responders, are ever-present but in a remote and backstage fashion. It was often difficult for paramedics and techs to establish a trusting and mutually beneficial relationship with these faraway colleagues. There were tensions between them, sometimes expressed in 'them and us' terms. Many on the road would say that the progress of a shift will depend very much on 'who's on Control that day'. Some of the dispatchers were portrayed as 'not using common sense' and having an unhelpful approach. Ambulance crews would often muse about their own personal ideas for 'fixing' these problems with 'Control' and 'management', feeding into a broader narrative that 'they' have 'forgotten what it is like out on the road'. On the other hand, some paramedics would point to examples of people working at EOC that they had 'got to know' via radio and telephone communications over the years. Some were described as friendly and very supportive. Others, however, were considered an unknown, incompetent or even malevolent presence. Developing trusted relationships with remote workers can be difficult. In stark contrast, paramedics greatly valued their connection to the more tangible and direct presence of their co-workers on the street and on stations. The camaraderie associated with working with other responders in the field is the subject of the next section.

Sirens and *Bringing Out the Dead*: Teamwork and the Personalities of Crewmembers

Operational paramedics interact with all manner of other staff when out on the road. Obviously the experience of having two people working together on an ambulance is extremely common. But beyond this, a crew will regularly arrive at an emergency scene where another crew is already present, and an impromptu team will form whose job it is to treat the patient(s) and 'manage' bystanders and family members. Often this is a collective effort at discretionary problem-solving (Lipsky, 2010; Maynard-Moody and Musheno, 2003), a classic example being the tricky task of safely extracting an immobile or partially mobile patient down flights of stairs. Paramedics will frequently come across crews from other stations elsewhere in the locality, or even ambulance staff from different trusts, especially at hospital A&E units located close to trust borders. Crews will also encounter professionals working for the police, and the fire and rescue service, as well as dozens of NHS occupations in and around various NHS facilities: doctors, nurses, midwives, cardiac technicians.

While solo responders sometimes experience loneliness and boredom, especially when sat on a standby post, for most ambulance crews interactions with others are pretty much constant. By far the most common relationship, and usually the most intense, is formed between 'crewmates' working together on a double-crewed ambulance (DCA). Many paramedics reflected on this aspect of their work in depth. Just as the prospects for 'a good shift' will largely depend on 'whoever's on Control that day', the same dynamic would apply to who your crewmate is for the next twelve hours or longer (see also Corman, 2017b). This is how Simon reflected on this dynamic:

Simon: Did you ever watch *Sirens*?[4] I thought *Sirens* was done pretty well. It was the only thing that I've seen that captures the weirdness of the work, the absurdity, the ridiculousness. The protagonist I can relate to. One moment he is sobbing, the next laughing hysterically. And the interplay is done quite well. The colleague you are crewing with—he or she defines what your day will be like. That part is also done well in *Bringing out the Dead*, that Scorsese film, have you seen that? He crews with three different guys and has totally different experiences—that is true, that is absolutely true. Some colleagues are brilliant and you are relieved to find out you are working with them on the roster. You interact and you both learn, you swap experience. You also bump into one other during shifts, turn to each other for advice while at A&E. We look after each other. Is my partner helping to keep me safe? I remember crewing with Richard, we'd been to a good job, turned it around, we're heading back out of the hospital, just

---

[4] *Sirens* was a British Channel 4 TV comedy series about ambulance services that typically featured two paramedics and a student observing. It is loosely based on the *Blood, Sweat and Tea* (Reynolds, 2009, 2010) blog books written by 'Tom Reynolds' (real name Brian Kellett).

cruising down the road with no job as of yet and he turned to me and said, 'this is fun, isn't it?' The person you are with makes it a different job.

There are also other ambulance-type organizations on the road; non-NHS crews doing very similar work to the paramedics and techs employed by NHS Trusts. This differs across the country, but in many regions of England you will see a small number of third-party operators that are contracted to NHS ambulance trusts to help them cover their areas and hit their performance targets. These include private, for-profit operators and charities such as St John Ambulance. These units are staffed by a paid crew, usually trained to ECA level. Community First Responders are also present across the country—unpaid volunteers trained by NHS ambulance trusts who use their own private cars to respond to ambulance callouts. NHS crews' opinions on these operators were interesting. Sometimes they would express criticism of the fact that these crews exist at all, and would discuss this as part of the stealth privatization of the NHS. But, when it came to working with them, their relationships tended to be supportive and helpful. Occasionally a crew might jokingly characterize CFRs as 'enthusiastic first-aiders', but deep down they welcomed the existence of extra pairs of hands on the road. I came across a St John's crew when observing on a shift with Emily on a solo vehicle. They provided onward transportation of a patient that we had arrived at first. I noted in my fieldnotes that they all worked together with ease:

> We head back out of the patient's house. The St John's crew gets the patient out on the evac-chair and they help her and her husband up into their ambulance. The tech who is driving talks to us through his cab window for a bit about jobs, demand, etc. He is relaxed and quite cheerful and friendly, but clearly they are also – like the NHS crews – very busy. It seems an odd situation. They have different employers, work out of different stations, have different uniforms and vehicles, but the patients probably don't notice or care. The crews work together easily and clearly know each other. In fact, it wouldn't surprise me if Emily knows them better than she would another NHS crew from a different region.
> 'Yeah, they're a really nice crew, I see them a lot,' says Emily as she gets back into the driver's seat and closes the door. She presses the buttons on the MDT to show that we are clear and back active.

As in any job, how one interacts with colleagues from across various occupations and of different levels of seniority does not go unnoticed by others in the workplace. People develop reputations. There are some paramedics, EMTs and ECSWs that are disliked around stations and localities. While complaining and moaning are near constant in the ambulance service, some staff are regarded as excessively grumpy and unfriendly, often described as 'dragging everyone else down' and 'a bad influence' on student learners. A few are considered arrogant; known to be dismissive of the input of non-paramedics and said to treat lesser-qualified crewmates as 'bag carriers'. Paramedics who come over as bossy, intrusive, and superior are unpopular. Other staff try to avoid crewing with them. As Simon intimated above, it is difficult to experience a good shift out on the road if crewed with one of these unpopular characters.

On the whole, I found paramedics to be considerate to others and highly socially skilled. With their sensitivities heightened by years of working with those who are suffering, ill, anxious and unhappy, they tend to be extremely good 'people watchers' and are well-attuned to both their own behavioural patterns and those of others. One aspect of interprofessional street practice that I found particularly interesting was the complex dynamic of informal switching and adapting of roles according to the unique clinical, emotional, and practical demands of each 'job', taking into consideration the abilities, confidence, and comfort of the various staff members present on scene, as well as the needs of patients and bystanders. Sensitive, collegiate paramedics would be careful not to upset a lesser-qualified crewmate by degrading their skills, while also being sensitive to the risk and embarrassment of potentially overstretching them by extending them the responsibility for a patient that was potentially beyond their clinical competence. Interested in classic anthropological research in other settings, I came to understand this as part of an informal 'code of the street' (Anderson, 2001) a sense in which everyone 'out on the road' was learning and figuring out 'how to act' among one another, balancing how best to serve the patients and family members, how best to encourage and develop other staff, and how best to work through each 'job' so that everyone gets to the end of the shift without endangering a patient or experiencing or provoking unnecessary conflicts or problems. This was a fascinating example of experiential-based behaviour, a form of workplace learning by doing, critically influenced by informal and unspoken signals given by various role models. I will return to this point in more detail in Chapter 6, as it speaks directly to that chapter's focus on professionalism and professional identity.

As I came to better understand the nature of paramedic work through my interviews and ride-along observations I increasingly noted the crucial importance of this informal, practice-based, and experiential learning, and the unstated yet critical ways in which this learning is shared and reproduced. As with any occupation or group, the technical skills that are formally taught, assessed, and certified are only one element of what makes a person 'a professional'. What is equally important—probably more important in a climate where certified technical abilities are a minimum requirement—is the process of learning how to act; the taking on of behaviours, developing into a skilled and respected practitioner who knows how to handle a scene, and can genuinely and convincingly embody the role of an effective responder and trusted colleague (Filstad, 2010; O'Toole and Calvard, 2019).

This was brilliantly described by Simon, in the way he talks about informal, unspoken routines around swapping driving and attending duties. This would often happen without anyone saying anything; it just somehow emerges through the interaction of behaviours and expectations that circulate people's various roles, physical locations, and duties. It can be manifest in something as subtle as the talk and gestures relating to the ambulance's ignition key:

Simon: There is a huge range of the kinds of person working in the support roles. There is a huge spectrum of ability. Some are ECAs who really should go on to do the paramedic qualifications. Others are just drivers and carriers and don't want to be anything more. I always like to work on the basis that we are sharing the jobs, working together. I think the patient profits more from this. It is unspoken, kind of informal. You just take turns to attend, one driving, the other attending, but if the job is too much for the ECA then he will come round and just ask for the keys. The patient doesn't really know the dynamic, the difference between us, and doesn't need to know. It is not set in stone, but it is the way we work. It is usually the ECA's decision who drives to the hospital. If the job is a bit beyond them, they will ask for the keys. Sometimes I'll step in if I think the job is beyond them. You know, I'll kind of hand him or her the keys—'I'll see to the patient this time, you can drive'. It is informally regimented, if that makes sense?

Simon's reference to 'a huge spectrum of ability' refers in this case to the different personalities, employment histories, skills, certification, and training and education backgrounds of the various people involved in pre-hospital ambulance work: ECSWs, EMTs, paramedic students, IHCD or degree-trained paramedics, supervisors, managers, and the occasional HEMS or BASICS[5] doctor. It is indeed a complex and varied field. But the 'spectrum of ability' also has resonance with the variety of skills, behaviours, and attributes embodied in different combinations by each person working out on the road; the individual styles of practice that professionals develop and come to embody. Paramedics have an ambiguous, hybrid, changing identity. An important element of ambulance work is its unpredictable nature as a form of uniformed, emergency work, and a reality that some of this work has an 'extreme' or 'edgework' character (Granter et al., 2019) that will always be a certain part of the work's attraction. But, fundamentally, paramedicine is a caring profession. Staff are professionally and legally bound to behave according to the high standards expected by patients, the public, ambulance trusts, the profession, and regulating bodies. A caring professional should be supportive to all persons they come across in the course of their work, including colleagues in other NHS services and one's own crewmates. Of course, behind the closed doors of an ambulance cab or an ambulance station it was common to encounter 'robust' views about patients and their behaviour, sometimes wandering into rather unkind and even 'extreme' forms of joking, bantering, and swapping of stories (Charman, 2013; Tangherlini, 1998, 2000). Private interactions between ambulance staff away from patients and managers would involve mutual support and care, informal advice and reassurance, as well as gossiping, venting, moaning, and posturing of all kinds (see Chapters 4 and 5). But at the forefront, when facing patients, bystanders and family members, the bearing of the paramedic is almost always one of care, attention, and respect. The complex interaction of paramedics with the public is explored in the next section.

'The Best Job in the World': Working among the Public

A paramedic's life is not easy. There are tremendous pressures and strains associated with working out on the road (see Chapter 5). But these demands can be often offset by the rewards of this calling (Granter et al., 2019). Rachel was one of several of my interviewees who described it as 'the best job in the world'. Emma spoke of the challenges involved in supervising frontline roadstaff who are perennially exhausted and frequently complain

[5] HEMS is Helicopter Emergency Medical Services. BASICS is British Association for Immediate Care, a charity that provides training and placements for doctors to work on emergency response cars.

of low morale. Supervising roadstaff means listening to a lot of complaints and frustrations. But she also indicated that there was something intrinsically good and rewarding about life as a paramedic that allowed her to 'always turn them around' to focus on its better sides:

> Emma: They moan for an hour and then say: 'it's the best job in the world'. In how many other jobs do you get to save lives? You don't do it often, but sometimes, yes, you've made a huge difference to people.

Carl expounded on this:

> Carl: We're popular with the public. The public likes us, what's not to like? We're seen as hardworking, professional, will help you in your hour of need. Media can be critical about us turning up late to jobs and that sort of thing, but the overall view the public has of us is generally very good.

And, similarly, Simon spoke glowingly about the respect that the role has among the public:

> Simon: I wear the name 'Paramedic' on my shoulder. I've always wanted that. I am proud of it. Ever since I became an ambulanceman I've wanted it. I like it when people ask me what I do and I say: 'I'm a paramedic', people are usually impressed. They look up to it to some degree.

In addition to the social esteem that a paramedic can enjoy, a vital element in making paramedic work satisfying and (at least partially) sustainable is the fact that ambulance staff usually face only one patient at a time. This is not the case for many types of NHS clinician, especially those working in acute hospitals or mental health trusts where doctors, nurses, and others have large ward rounds and are handling the care of ten or more patients pretty much concurrently.[6] The ability to focus on one patient at a time is an important part of the 'privilege' of being a paramedic that I mentioned at the opening of this chapter. NHS staff of all kinds mostly support the vision of the NHS as 'practical socialism' (Player and Barbour-Might, 1988),

---

[6] To be more precise, having only one patient is overwhelmingly the norm, but emergency calls that require attendance to more than one patient do, of course, happen, and can be very challenging to handle. These might range from domestic and street disturbances to multiple-vehicle collisions. There is also the possibility (albeit much rarer) of mass-casualty incidents (train crashes, terror attacks). Incidents of this kind can be overwhelming, potentially leading to long-term psychological injury for the responders (see Chapter 5).

a service that provides care 'free at the point of use' that is available to all persons irrespective of their ability to pay. The ambulance service embodies this ideal very strongly—or at least that's the aim. Paramedics are widely available and are expected to attempt to resolve problems for patients in a non-judgemental fashion, focusing on each patient's medical problem, whatever that is and however it came about. They are trained to make good communication with patients and family members central to their practice (Willis, 2015). They are encouraged to take patients' descriptions of their complaints seriously, right up to the point where the 'pain score' of 1–10 is whatever the patient tells them it is. I came across many examples of just this kind of care in my observations. Crews would be careful and sensitive. Care would be patient-centred. Along with clinical discussions (symptoms and complaints, mechanisms of injury, medical history), they would usually listen sympathetically to the patients' concerns and fears, and would try to improve the mood on scene with a light-hearted comment or observation, usually involving the kind of self-deprecating humour that they often excel at. They would build rapport with patients and try to relax them by asking about their background, their families, their hobbies, and activities; I found this to be especially common with calls involving children or elderly patients. At calls at patients' homes, paramedics and other road staff would sometimes talk through the photos of family members hung on the wall or perched on the fireplace. Many paramedics are highly skilled at 'managing the scene', adjusting their emotions and influencing the emotions of patients and others present (Campeau, 2009, 2011).

Of course, they do form judgements about patients. Having seen so many patients facing critical distress and death, paramedics can hardly avoid developing private judgements according to an experiential calculus of patient 'need'. Often this would be connected to simple issues of practicality—to what extent are the 'needs' of the patient proportionate to the efforts being asked of the responding crew? An ambulance crew would sometimes be quite forceful in encouraging a patient to walk themselves downstairs and out to the ambulance even when the patient would try to insist on being carried out using an evac-chair or similar device. Often there were suspicions that a patient is exaggerating the severity of their condition. Occasionally there would be a peculiar extension of the 'drama' of an ambulance callout, whereby a patient might be 'faking it' or 'playing up'. Several times I witnessed ambulance crews criticizing patients' behaviour, demeanour, and assumed way of living, but only ever in private. Sometimes this would be done very subtly, with a knowing comment or a quiet and obscured gesture such as an eye-roll or a pulled face behind the backs of patients or family members. Often an entire patient episode would be criticized by the crew as pointless and undeserving, feeding into the narrative that 'Control' has no sensible grasp of call prioritizing. The following field-note describes a situation where scarce ambulance resources were indeed 'wasted' on a patient who did not need a callout:

We are in a town centre, about two minutes into a blue-light drive to a 76YOM, DIB. Further information on the MDT suggests the patient is very sick indeed, with late-stage cancer. But for some reason we are cancelled down from that call and immediate get sent another—'42YOM, seems confused, shaky. Query stroke.'

Chris: 'It's a Cat A, Red 2, but it's 16 miles away. If we drive 70-80 miles an hour with no traffic we'll still miss it.'

The call takes us into rural areas. We rush past fields and woodlands, at some points moving at speed down narrow, winding country lanes. Out in the countryside I start to wonder how well-prepared oncoming motorists are for large emergency vehicles heading in the other direction. Preconditioned by 'reality' TV, somehow the wailing ambulance seems to me less suited to the rustic lanes of England. One tends to think of the rushing ambulance in the context of clogged road junctions, blue lights rebounding around the concrete jungle, pedestrians thinking twice before crossing the road. But here we are, rushing through pretty countryside, the soundwaves of the siren blasting through trees and bushes. We eventually come to a turnoff down what is little more than a gravel track sloping downhill. At the end of the track is a five-bar gate and an RRV parked next to a small passenger car.

We come to a stop as the radio comes alive again—'there's an issue with the patient' - and this is taken to mean he might be a physical risk to the crews. Chris gets on the radio to clarify. While this is going on I note that the RRV paramedic already on scene is talking to a young woman. It doesn't look dangerous at all. We head outside and walk over. The paramedic (Mike) is talking to a woman in her mid-late twenties in smart clothes who is worried about the patient and had called 999. It turns out the 42 YOM is a friend of hers who is staying in some kind of temporary accommodation; a very small bungalow overlooking some fields. Sat at an outside table is the patient. He is leaning over, looking unwell and in a very bad mood.

'Ah, this virus is fucking killing me, man.' He looks in a bad state, like someone living on the streets.

Mike explains that the young woman found him lying on the ground outside. Understandably she was worried about him and called 999. Mike has taken

basic obs and there is nothing obviously wrong with him.
Mike asks the patient a few questions to which he doesn't receive especially helpful responses. 'I just want to sleep. Let me sleep.'
We've all noticed an empty bottle of vodka on the table, plus a full ashtray and a lot of torn cigarettes, like he's been pulling the tobacco out for rolling marijuana joints. It might be a case of being drunk and ill.
'It's just a virus man, it's a fucking horrible one, I feel horrible, man. But it is ok, I don't need any help.'
Mike tries to ring the man's GP, based in a suburban area quite far from here. The man says he is 'visiting' and staying here. The place is a wreck inside, booze bottles, clothes strewn across the floor.
They ask him how long he's been ill, take basic obs, etc. Temperature is around 35-37, quite normal. They can't get through to the GP. Reception is closed.
I notice a teaspoon that is burnt and contains the remnants of a very suspicious looking substance. Sally asks the patient about this.
'No, no ma'am, it's not drugs, I've been eating yoghurt out here. I don't do drugs, none of that business,'
'It's alright, you're not in trouble! But have you taken anything, used any drugs, alcohol?'
'No, I promise you. No, it's just this virus. I need to go to sleep, I need to rest, that is all. This virus, its fucking destroyed me, man.'
Eventually the crews decide to leave him there. He has mental capacity and does not want to go anywhere. Mike and Chris agree that A&E would be pointless. So the paperwork is filled in, and the patient signs the disclaimer that he has decided to stay at home. Under 'Diagnosis / Condition' Mike puts in 'General Malaise'. 'Ah he's good, the General,' he says. We joke about this a little. 'Yes indeed, a good tactician, a great leader!'
We eventually head back to the vehicles. The crews ruminate about why the young woman would want to have anything to do with that man. 'What a nice girl,' says Mike, 'Driven all the way down from work in Charterville to look after that idiot.'
'I don't know where they got 'stroke' from' says Chris. There was no way that was a Red 2. I myself wonder why we were summoned to that 'rubbish' job on blue lights when no-one was in danger.

The above is a good example of roadstaff behaving professionally in public, while being critical of the patient in private. The experience of being diverted to that call from an elderly cancer patient seems especially jarring, but maybe there were good reasons for this which the crew will never learn about. These scenarios vividly indicate the disjuncture that can erupt be-tween crews' experiences of the road and Control's understanding of call priorities.

On the other hand, as a flipside to all the talk of 'unworthy' patient episodes, paramedics also discussed other patients' severe social disloca-tion and vulnerability, expressing compassion and empathy as they did so. Calls to vulnerable communities represent a major proportion of the work of the ambulance service, the NHS more broadly, as well as social services, police, and the courts system. A young paramedic named David de-scribed some of this to me in an interview. Highly articulate, David talked in depth about the people who had, in his words, become 'the new untouchables':

> David: There are kind of tiers of homeless people; how far they've fallen. The ones right at the bottom of the heap are not really a problem for us. They never actually call us themselves. Sometimes someone will call on their behalf. But their own sense of self-worth is so low, their own standards of health and wel-fare so low, that they ignore their own problems. Occasionally we will go out to them. They often smell very bad and you have to get the ventilation on in the ambulance. You try not to gag. You just do your best. Sometimes you have to wrap them up tightly in blankets to try to contain the mess and smell. But then what do you do? You often see them wandering around the hospital corri-dors, walking out, that sort of thing. In a way it is not really our problem. They are so neglectful of themselves. In a way they are the new untouchables of our society. Many of them have major psychiatric problems, PTSD and similar is-sues. I remember a call in Crantonville where the patient had reportedly taken speedball—a combination of heroin and cocaine. He'd OD'd and this was poten-tially life-threatening. We were there to help him, but the issue was more to do with the person he was with at the time. She was known to the emergency ser-vices, she is a vile creature, abusive and manipulative. So we had to find a way to keep her away from what we were doing. We gave the patient Narcan (Naloxone) to reverse the effects of the opiates. Once in the ambulance we then realised he was crawling with lice! It was horrendous. I'm laughing about it now, but it's just ridiculous that people can get into this kind of a state. Really sad. He was effec-tively a public health hazard. We had to put more protective clothing on, try to make sure the lice don't get on your arms and clothes.

In a way that was quite common, having been explicitly critical of mem-bers of the public, describing one as 'vile' and then involuntarily laughing at the absurdity and tragedy of it all, he then went on to check himself, describing the broader context of poverty that puts so many people into such precarious and vulnerable positions:

> David: It's getting to be a severe problem—there is this little tent city down in the middle of that roundabout, just near Upperton where I live. I walk past it often. It is sad to see people whose lives have such little dignity and self-respect.

This is a good example of the way in which the locus of the paramedics' professional bearing nearly always circulates back to care for the patient, even amid the often uncompromising and traditionally masculine realm of 'street work' (Kyed, 2019). Other fieldnotes describe patients facing a mixture of chronic physical and social distress. The following is an example of quite an everyday callout, in which the solo paramedic responder did all she could for a chronic illness sufferer, but within a broader operational structure that leaves the immediate and longer-term outcomes unresolved and unknown to the responding ambulance worker. Paramedics and other ambulance professionals rarely get a sense of closure about their patients, a facet of their work that sometimes contributes to feelings of frustration, alienation, and even low self-worth among the emergency community. The following fieldnote gives an indication of how inconclusive ambulance work can be when many of the calls emerge from chronic medical, social, and psychiatric problems, rather than from sudden, acute emergencies:

> The radio bleeps. We get our first call at 0830. It's a 111 redirection[7] from a residential care home. MDT says 'chest and arm pain' which sounds ominous, but upon arrival it is immediately obvious that this is not a life-threatening emergency. We knock on the door and an elderly lady opens. Her home is very small, there is a tiny bathroom near the door, a reasonable-sized living room, a small kitchen and a very small bedroom.
>
> The patient is very agitated, and her behaviour is odd. She is pacing around and it is hard to get her to settle so that Katie can do a patient assessment. She won't really listen and to my novice eye Katie's demeanour is somewhat hard with her, rather than 'caring' and compassionate. The more time we spend at the scene, however, the more it dawns on me that this is a mental health rather than an internal health call, and that Katie feels she has to be 'firm' with her to get anywhere with this job. She can't remain with this patient for hours on end.

---

[7] 111 is the NHS's 24-hour telephone advice service that is intended for non-emergency health issues. Call handlers often escalate 111 calls to the ambulance service where potentially significant risks to the patient are identified (see Chapter 6).

Katie is trying to establish herself as in charge. She ushers the patient into the bedroom and keeps telling her to lie down on the bed so that she can do some checks.

The patient keeps gasping and saying 'oh my god, what am I going to do?' and 'don't take me to hospital again, they'll do nothing.' 'Please don't leave.' *Please do something*. When Katie goes back to the bags to get the temperature monitor, phone, or radio, or to look at the various care home paperwork, the patient keeps calling out to her, calling her 'nurse.' She is complaining about pain in her abdomen and says that she wants to go to the toilet but can't. Katie is unflappable and takes an ECG reading which is completely normal. Oxygen saturation and temperature are also ok—there is nothing seriously wrong with her in a physical sense. It is probably just constipation and an acute anxiety attack. She really needs to lie down and take some deep breaths.

Katie says quietly to me: 'I don't know where they got chest pain from. I'm not going to ask for a pain score as I know she'll give me a 'ten'.'

'Please don't take me to hospital again – I can't stand it. My grandchildren are scattered all over the place. No-one will help me. No-one cares about me.'

Katie is a Paramedic Practitioner whose role is supposed to be about precisely these kinds of social and unplanned primary care calls, with the aim that treating people in the community keeps them away from hospital. She spent about an hour at that call, making inconclusive telephone calls to GPs, family members and the care home management, and rummaging through unclear paperwork. In the end the patient was transported to the A&E in any case, as there was no other way to 'resolve' it. Katie expressed frustration that this patient's mental health needs were not being adequately treated at the care home, and that a trip to A&E, while frustrating and inappropriate in many ways, might at least lead to the patient being given a psychiatric care plan that she seems to need.

The above fieldnote described a rather sad callout. It was the kind that prompts the observer to wonder about a patient's immediate and longer-term prospects, given that it is questionable what a local A&E department can realistically do for patients with chronic mental health and social care conditions. For me at least, it served as a useful lesson on the feelings of uncertainty and inadequacy that an ambulance responder might experience. Lipsky (2010/1980) teaches us that a street-level bureaucrat does whatever he or she can for a client in a constrained and contradictory operating environment. That is certainly true. Calls like the above call into question the design, value, and distinct limits of a 999 'emergency' system in trying to confront such protracted features of social distress.

Conclusion

This chapter has provided a preliminary overview of the working world of an NHS paramedic. What I have aimed to illuminate most of all is the complexity and range of the work. While the work is structured in different ways by the employer, involving a range of base stations, shifts, vehicles, protocols, geographies and communities, the fundamental nature of paramedic emergency response has certain core features. While the paramedic crews have little to no control over which calls they are sent to, the actual work of responding to calls requires a great deal of operator discretion, skill, adaptation, and improvisation, very much in keeping with the experiences of other street-level bureaucrats (Lipsky, 2010/1980; Maynard-Moody and Museno, 2003; Vinzant and Crothers, 1998; Riccucci, 2005). Although a paramedic can often feel isolated, especially when working on a solo response vehicle, the work often features teamwork, and relies on cooperation and mutual support, amid very strong occupational norms (Charman, 2013; Seim, 2020). The specific work tasks performed by paramedics range from technical (Barley and Orr, 1997; Orr, 1996, 2006), through to emotional labour and emotional management (Boyle and Healey, 2003; Filstad, 2010). The work is highly mobile, yet there is a management presence that—even when barely audible and rarely visible—encapsulates the entire work setting, rather like a computer operating system buzzing away in the background. This operating system prioritizes and dispatches the calls and evaluates work performance, and its remote and often depersonalized nature encourages paramedics to treat it with suspicion. Complaints about management structures, styles and personnel were a common feature, often expressed via a 'them and us' narrative that 'management' has 'lost touch with the realities' of the road.

Subsequent chapters will go into further depth about the nature of the paramedics' occupational culture. As we shall see in the next chapter, the distinct cultures and structures of NHS management and organization play critical roles in establishing the parameters around which paramedic professionalism can either express itself, sustain itself, or even struggle to establish itself at all.

4

The Operator Culture

> 'The uniform acts as a guarantee that an upper level in the group
> will control the members, and in turn, that members will conform.'
> **(Joseph and Alex, 1972: 723)**

Paramedics are routinely exposed to physical and mental exhaustion, to emotional trauma and to medico-legal risk. Shift work is not family friendly. The pay isn't great. There are easier ways to make a living. Why would someone join the ambulance service? I asked this question of many of the paramedics I interviewed and accompanied on shifts. One of them was Becca.[1] I asked her this question while she was driving back towards the station having dropped off a patient. Her answer:

> This sounds corny but it's true. I was a student at Campuston University. I
> thought the course was shit and it wasn't for me. I remember seeing an
> ambulance go past me on lights and sirens and the crew were really laughing
> about something on the way. You don't really know what you're looking at, but
> I remember thinking, 'that looks wicked'. I had imagined it would be a constant
> thrill. Can it really be like that? I wanted to try it.

Becca was one of the more assertive characters I came across. She came as close as anyone to conforming with Palmer's (1983) notion of the 'trauma junkie'. A Critical Care Paramedic, she would expect (and hope) to be dispatched mostly to clinically very serious calls, often involving danger to life. A day spent observing on her vehicle was full of bravado and banter. Larger than life, she was at home in her role and was widely respected by others on the road. She could be disarmingly direct in sharing some bizarre and ghastly stories. Off duty she was into extreme sports. But the 'trauma junkie' could only ever be a part of her identity. Amid the enthusiasm and energy, she also expressed misgivings and frustrations about the purpose of her role, and of the ambulance service more generally. Although always

---

[1] All names of people, places and organizations are pseudonyms. For details on the research process involved with the writing of this book, please see the Appendix.

The Paramedic at Work. Leo McCann, Oxford University Press.
© Leo McCann (2022). DOI: 10.1093/oso/9780198816362.003.0004

giving the impression of being totally unfazed by the death and destruction that often came her way, her behaviour and bearing also featured examples of kindness, care, introspection, and philosophizing.

A person is always made from an assortment of characteristics. The same is true for an organization or occupation. The paramedic profession sometimes seems especially complex and schizoid. Its activities, its patients, and its emergency scenes are as varied as the characters who wear the uniform. While a uniformed occupation will feature certain aspects of conformity (Joseph and Alex, 1972), there is no paramedic 'work culture' in a singular and defined sense. And yet 'culture' is a vitally important feature in discussions of the nature of the paramedic profession and of ambulance services as employers. A multifaceted 'ambulance service culture' was readily referred to by the paramedics I observed and interviewed in this study; a culture that features a broad array of characteristics both attractive and unpleasant.

This culture is quite particular, made of constituent parts drawn from a range of sources: employment relations, educational backgrounds, notions of professionalism, and the varied and complex personalities of the individuals who work in the service. Like a volatile mixture of chemical gases, the reactions and interactions of these cultural traits are compressed and intensified by relentless operational pressure. The drama that sometimes accompanies emergency scenes can add a catalyst that accelerates and complicates these reactions. Interpersonal relations between paramedics and other roadstaff were usually highly supportive. But conflict and tension were ever-present at the edges, particularly at what might be considered the 'frontiers' of management control (Hughes et al., 2020), such as the relationship between crews and dispatchers at EOCs, and between paramedics and their local and regional line management.

This chapter focuses on what has been described as the paramedics' 'operator culture' (Wankhade, 2012), one subset of the broader 'uniformed' occupational culture associated with other emergency response professionals (Charman, 2013; Chetkovich, 1997; Loftus, 2010; Maynard-Moody and Musheno, 2003; Moskos, 2009; Reiner, 2010: 115–38). It aims to provide an in-depth account of the working culture of paramedics, crafting a detailed picture of an occupation that shares some characteristics with other uniformed occupations such as policing and firefighting, but which has received comparatively much less attention in sociological research. Uniformed professions serve society and are part of society, yet have a degree of secrecy and isolation from it and a powerful sense of mystique

(Coser, 1974; Soeters, 2018). Part of this stems from the 'powers' that the uniform affords, the meanings it transmits to the public, and the burdens it places on the wearer (Joseph and Alex, 1972). While this book is about a profession, the fact that the paramedics featuring in it were all employed by NHS trusts means that the book is also about work, employment, and occupational culture. Paramedics enjoy a large degree of autonomy, but they also work in and around an operational system constructed by management and upon which management tries to increase its scrutiny and control.

The English ambulance service culture is a hybrid of a uniformed, emergency service culture, and a more 'NHS' type clinical culture that is recognizably similar to those existing in other health organizations such as hospitals, GPs surgeries, or mental health trusts. The 'emergency services' side of the culture is reflected in ambulance Trusts' retention of a rank structure, complete with pips or cartwheels on the uniform epaulettes. Some Trusts have considered removing or changing the rank structure, seeing it as out of step with the clinical culture of the NHS and a barrier to a more open, white-collar, and perhaps even corporate, culture. But these change attempts often hit a barrier from operational staff (usually long serving) who continue to argue that rank and insignia will always be necessary because of the need to know 'who is in charge on scene'. Unlike the rest of the NHS, ambulance staff share an affinity with other emergency services, especially with police, who they interact with quite regularly and who provide rapid backup if ambulance staff press a button on the radio asking for 'immediate assistance on scene'. Technical and quasi-military terminology such as 'stand down' and 'return to base' circulates ambulance work, especially on radio and MDT systems which make up the most direct control interface between 'management' and 'roadstaff'.

The traditional management structure of ambulance services also reflects the 'emergency services' culture. The traditional way of managing is a 'command and control' mentality, based on rank, doctrine, and hegemonic masculinity, an approach now widely seen as limited and outdated (see Caless, 2011; Charman, 2017; Holdaway, 2017; Turnbull and Wass, 2015). The weaknesses of this kind of approach (Grint, 2010, 2020), is of particular concern here as the ambulance service is progressively moving away from its roots as an emergency service focused on responding to trauma as it evolves into something closer to unplanned and mobile primary care. But there are also plenty of traditionalists who continue to favour 'rank' and 'command' approaches. Managers are occasionally referred to as 'ambulance officers'. Frontline responders have traditionally developed an embodied, uniformed identity that matches the uncompromising realities of the work they are entrusted to carry out. Uniformed cultures can be highly change resistant (Courpasson and Monties, 2017).

Such discussions recall classical Weberian notions of vocation, calling, and 'office'. These notions leverage an understanding of work and duty in which operational professionals don't need managers because they know more about their field than a general manager ever could, and are able to build and defend their knowledge base to gain occupational closure. Professionals might be reluctant to accept being 'managed' by persons from outside the profession, and a non-expert manager can feel uncomfortable giving instructions to experts (de Bruijn, 2010). This might be especially the case with uniformed services, where distinctive powers may be invested in the role and where misconduct in a public office can be an offence carrying heavy penalties.

Things become yet more complex when one considers that emergency professionals tend to be employed as salaried workers in hierarchical organizations with multiple management layers. Many NHS, police, fire, or ambulance managers are themselves uniformed professionals in hybrid professional/manager/officer/operational roles (Currie et al., 2016; Hyde et al., 2016). How can an organization manage autonomous professionals? To what extent should it try? Can senior roles be adequately filled by persons who have not climbed the internal uniformed ranks or who are not members of a profession where their registration is 'on the line' if things go wrong?

Many existing studies of police, firefighting, and ambulance work use in-depth, qualitative, and ethnographic fieldwork, often involving researchers immersing themselves into the mysterious, sometimes intoxicating culture of emergency work. This occupational setting exposes workers to society at its most raw and unvarnished, providing rich sociological material for researchers. Certainly, the particularistic and eye-catching elements of emergency work (risk, hierarchy, camaraderie, symbols of authority) are important parts of the fairly distinct social roles that uniformed occupations represent (Joseph and Alex, 1972; Loftus, 2010). But there are other stories to tell about emergency services. The one I tell here describes how the occupational norms of paramedicine are changing, gradually coalescing around the normative power of the concept of 'professionalism' (McCann and Granter, 2019). I will explore how the traditional forms of uniformed culture—such as Donald Metz's (1981) classic notion of the ambulance worker as 'blue-collar professional'—are evolving into something new. The new culture is not quite fully formed, not fully broken from the old. Strong remnants of traditional uniformed traits remain, including norms around hierarchy and time served, a sense of hegemonic masculinity, and a robust defence of street-level craft and discretion learnt not in a classroom but from experiential learning on the road.

The chapter proceeds by exploring five interlocking aspects of operator culture. The first two sections discuss largely 'operational' issues surrounding paramedic work: how paramedics account for and explain the autonomy afforded to them in their roles; and the influence of managerial targets and performance indicators on their work. The subsequent three sections move on to explore more specifically 'cultural' manifestations of this work climate: the informal value hierarchy that paramedics ascribe to the kinds of 'job' they respond to; the ways in which they interpret, discuss, joke, and complain about their work; and the changing composition of the workforce, featuring rapid change in gender relations but very limited ethnic diversity. Taken together, we have an interesting intersection of themes. The chapter tells a story of how a blue-collar culture is increasingly taking on the trappings of 'a profession' yet is doing so within an employment context and operational model that are troublesome, conflictual, and exhibit tendencies towards maintaining certain traditional elements of occupation culture that do not always sit easily with classical expectations of professionalism.

We begin with the vital issue of operator autonomy.

'You and Me against the World': Paramedics and Autonomy

As discussed in Chapter 2, one of the central elements distinguishing a 'professional' occupation is the large degree of discretion and leeway that comes with the role. This is a complex area of discussion, because it is also true that many lower status 'technical' occupations also provide discretion and allow workers to operate with limited supervision (Barley and Orr, 1997). Work that is mobile in nature often possesses such features. Many occupations require workers to operate in remote locations, away from an office space or a retail unit where management is co-present, or where movements and actions are meticulously logged and monitored by punch-clocks, attendance registers, team meetings, computer systems, video cameras, and the like. The nature of work for ambulance road staff—in common with the work of patrolling police officers—very much fits Freidman's (1970: 136) notion of 'discretion by default' and Lipsky's (2010) 'street-level bureaucracy'. Managers cannot closely observe and micromanage paramedics' work out on the road. With crews typically leaving the station moments after having 'booked on', supervisors and managers often claim that the staff are then 'out of the door' and 'you never get them back' (Hyde et al., 2016: 74–105). A great deal of the day-to-day activities of ambulance crews has to be taken on trust. This is in distinct contrast to fire and rescue crews, where the work structure is almost militaristic, with fire appliances tightly 'crewed', always working as a close team unit, and with a ranking person in command of the vehicle and the scene. The quotation from sociologists Nathan Joseph and Nicholas Alex (1972) at the head of this chapter holds true in some sense, in that it speaks to a hierarchy of control in the ambulance service. But it should not be interpreted to mean that the street-level responders' wearing of a green uniform and steel-capped boots automatically entails a precise set of behaviours or universal conformity. As we shall see, paramedics have considerable discretion around how they operate in the field, although bounded in important and powerful ways.

In addition to the autonomy that mobility provides, paramedics are also taking on steadily growing levels of clinical autonomy, as their scope of practice has expanded over recent years (Campeau, 2016; Eaton et al., 2018; Newton et al., 2020). Paramedic graduates from university degrees are encouraged to act as autonomous professionals, whose clinical licence to practice depends on the rights and responsibilities afforded by their HCPC registration. Paramedics often note that they carry considerably more clinical autonomy and risk than other NHS and NHS-related occupations, such as nurses or other AHPs. This growing clinical responsibility (and the concomitant increased degree of risk it poses to the licence-holder) is largely welcomed by the paramedics, on the basis that paramedics typically want to be allowed to provide the broadest, deepest, and most flexible forms of care for patients that can be safely delivered.

This bearing was by no means exclusive to the degree-trained paramedics. Simon, an experienced IHCD-trained paramedic, had an interesting interpretation of what autonomy meant to him. His line was quite assertive, perhaps exaggerated at times. His narrative had a pro-patient, pro-paramedic, and anti-management feel. 'I'm my own boss', he claimed, with 'colossal autonomy':

Simon: When you're on an ambulance, crewing with someone, it's just me and you against world. It's sort of the same with a patient—you two and the patient against the world—'what can we do for you, how do we get you sorted out with the best care we can?' I'm my own boss. [. . .] [N]o-one really cares too much about how I work and what decisions I make about what to do with the patient. I feel we've got colossal autonomy. Between pressing 'At Scene' and going green, I can do what I want. Pathways exist of course, but we also make up our own pathways. A lot of what we do is improvised. Many times A&E is just not what the patient needs.

Simon exuded pride in the paramedic role, taking a strong position on paramedic abilities, roles and discretion. But his peculiar use of the word 'colossal' hinted that the degree of discretion could be a challenge. Might he have too much discretion? Lingering anti-management sentiment— strong throughout this study—came to the fore. What he was describing was not so much being granted trust and autonomy by an enlightened employer. Rather, the account he gave suggested staff were neglected by a management that he depicts as asleep at the wheel:

Simon: I've sometimes taken a patient to their home. I was once at a job where a guy was fitting on a bus. He was actually going to the football. I asked him how he was feeling, he had recovered well, he wasn't post-ictal, his obs were all good. He was just tired and worn out by it. I think also he had learning difficulties. But as far as I was concerned he was fit to carry on his journey, he could have gone to the game if he had wanted to. But he said, 'Really, I want to go home', so we took him in the ambulance and took him home. Why not? What's the problem with that? No-one checks. No-one checks because no-one cares. I'm autonomous. When I'm on that ambulance I'm basically my own boss. An autonomous professional.

This is a somewhat problematic description of paramedic autonomy. Acting as an autonomous clinical professional should not be synonymous with 'no-one checks because no-one cares'. Simon's portrayal of working relationships is not quite what is typically associated with a 'professional' environment. A culture in which professionals are employed should be one where the organization respects, supports, and rewards professional autonomy, but it should also involve time and room for training, for detailed debriefs, and for learning and peer support. I saw precious little of this in the ambulance service, at least not in a formal sense, budgeted and provided for by the employers. Sometimes the paramedics talked and acted in ways that seemed to downgrade the risks to patients and to themselves by the autonomy they had. I was sometimes surprised and troubled by this, but this was surely a reflection of my very limited appreciation of the experience they had built up by responding to thousands of calls over ten or even twenty years.

Paramedics often portrayed paramedicine as a neglected and forgotten profession, with large degrees of discretion and responsibility but without commensurate support, status, and recognition (McCann et al., 2013). Reflecting decades of inter-occupational rivalry—perhaps with echoes of the 1989–90 'coach and horses' dispute—comparisons to nurses and firefighters were common:

> Karen: Compared to Fire [and Rescue Service], they do things in groups. We work separately. Nurses start on Band 5 like us, but they normally have to check with the doctors and consultants every time they give drugs. We have the discretion to give drugs without this kind of clearance, we are on our own with the patient, there is no-one to check with. We have to get on with it. We now have about 40 different drugs we can give. We deserve Band 6. But who's gonna pay for it? Unison is going to issue strike ballots on the Band 6 campaign. We think we deserve it. Some trusts do pay it, we don't here in [. . .]. So, we'll see how that goes.

Simon made a similar comparison:

> Simon: I was in some in-hospital training a while ago. I was Band 5, almost everyone else was Band 7. But there were loads of things that I do out on the road that they are not allowed to do! Some of the decisions I make are far more clinically involving. We're very underpaid for the autonomy that we have, relative to others in the NHS. [. . .] If there is something the patient needs and I'm allowed to do it, I just do it, it doesn't need a whole load of checking and confirming with others. I sometimes think it's because we are largely solo operators—like a dentist. At that dentist's practice, he is alone, he is in charge. He will have his dental technician, his stooge if you like, but they don't follow all that the dentist is doing, they just do very basic work, they won't influence the clinical decision the dentists will make. I pride myself on making brave decisions when I have to.

Listening to a paramedic like Simon, one might get the impression that paramedics face no management control at all. That clearly isn't the case. Alongside the considerable freedom that ambulance crews operate with, they are also controlled in complex ways by various management policies, procedures and technical systems (see Corman, 2017b). These include: real-time locational tracking of vehicles, via radio, cellphone, and MDT systems; NHS trusts' line management procedures, policies and duty rosters; fairly strictly confined patient record forms; and a system of professional registration checks run by the HCPC. There is also the ever-present background possibility of management sanctions for poor performance, sustained sickness absence and other disciplinary issues. Management certainly does take notice if it learns that the condition of a patient discharged by a paramedic subsequently deteriorated.

It might be useful to provide a practical illumination. Image 4.1 is of the first page of a Patient Record Form or PRF—one of the most ubiquitous and everyday pieces of paperwork recently in use at London Ambulance Service. This document shows how managers and senior clinicians have attempted to design and construct the work of paramedics as inscribed in these documents. These documents should be completed during or soon after each patient episode. Clearly visible is the emphasis on monitoring the patient's medical condition and creating a paper trail which documents what the paramedic decided to do, what pathway the patient was taken through, which treatments and drugs were administered, and even which transportation tools were used to extract or move the patient. The logics of emergency and trauma dominate these forms, which is out of kilter with the new reality that ambulance work is predominantly now a form of unplanned primary care. Forms of this kind reflect the ways in which clinical pathways have been designed by other professionals and then 'laid down' for paramedics and EMTs to follow, much like the cardiac protocols described in Nelson (1997: 165–6). Subsequent pages of the form also include a legalistic section about decisions to leave a patient at home (including time stamps), where the form is designed seemingly to provide a narrative about each patient episode in case of complaints and legal redress.

The paperwork is somewhat restrictive although it does not fully try to 'force' the paramedic down certain branches of a tree of knowledge. At least one part of the form can be completed using free text which paramedics can use to document contextual details.

Although paramedics are formally and informally developing as autonomous professionals, the PRFs highlight broad and ongoing debates in the occupation about how management will try to structure, control, and monitor their work. A saying that circulates the paramedic world and an idea regularly reinforced in paramedic education is 'if it is not written

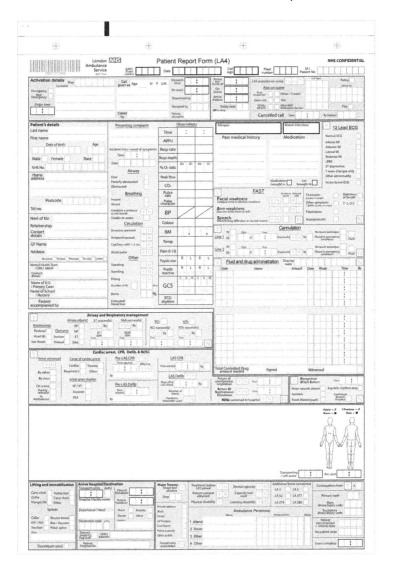

Fig. 4.1 First page of a manual Patient Record Form formerly used at London Ambulance Service.
Source: https://www.whatdotheyknow.com/request/london_ambulance_service_patient (accessed 21 January, 2022).

down then it did not happen'. This phrase is common across many public sector occupations where interactions with the public carry a degree of risk. It reveals occupation-wide assumptions about the thought processes of managers and (perhaps more importantly) regulators and inspectors.

The shift to 'new professionalism' (Evetts, 2011) mandates a 'professional' approach to paperwork in which 'professionals' are trained to take detailed notes of all actions, to dutifully log all interactions, to be explicit as possible in the paperwork, and to move away from old-fashioned and impenetrable occupational norms where staff would 'keep things vague' to allow discretion and leeway. The autonomy of new professions is clearly constrained by these procedures. Other systems also exist that attempt to 'format' paramedics' work. These are often established by other professions with higher clinical status, such as NEWS2 (the National Early Warning Score version 2), a risk management tool developed by the Royal College of Physicians that sets out to standardize paramedics' (and other AHPs') calculation of a patient's clinical risk, to better appreciate which patients are in danger of deterioration, and to inform decisions of which patients to bring to A&E. Paramedics operate under the control and scrutiny of employers andregulators, and regularly make use of guidelines and protocols laid down by other professional groups. But this isn't the end of the story. We know from the SLB literature that no form of paperwork, system or decision tool will ever capture all elements of the complexity of public service interaction (Brodkin, 2011; Maynard-Moody and Musheno, 2003; Zacka, 2017). Strictly following the rules of every procedure and taking the time to be fully explicit on all systems monitoring the activity of paramedics is not always desirable or even possible. Logging every detail about treatment information might also reinforce practices of 'defensive medicine' where clinicians become excessively conservative and risk averse.

At the time of my field research, PRFs were largely paper based; a block of forms held on a clipboard that the ambulance crew would carry to all calls, complete during each patient episode, and file at the station at the end of a shift. The paper forms were starting to be replaced by electronic Patient Record Forms (ePRFs) that run on a rugged laptop held on a detachable mount in the vehicle. A paramedic named Jamie complained to me about their introduction, regarding the rollout of the system as another example of management's rather remote and careless modus operandi, and how the ePRFs continue to reinforce outdated ideas of the paramedic as trauma responder:

Jamie: Things aren't helped by management issues. The new ePRFs – we were given one day of training for them. There has been a lot of issues with the technology and how it is set up. It's a [brand], quite a nice laptop, but the software on it has issues. It seems to be another system that is making us spend even longer at scenes. There was a meeting to discuss the trialing of the equipment and in the way it was delivered there was tremendous confirmation bias. Management had said 'this is what we're moving to, it's great', and any issues that bubbled up from the people actually using the system were just sidelined, basically, and talked away. Many of the regions we work in have poor mobile coverage. The decisions it makes you make, the decisions it sends you down, I call it Schrodinger's Paramedic! The way the menus and systems are set up is often not very helpful. They said 'it was a training issue', but it clearly isn't. [. . .] The system isn't up to it. It assumes that all emergency jobs are traumas. There should be four separate entry points to all jobs, something like: trauma, obs and gynae, psychiatric, dead. There is this mental inertia with our management. I'll go back to paper PRFs, thank you very much. The system is too rigid, the options on the screen are rigid. It's all set up to preset criteria: give it what it wants, anything outside of this it doesn't want to know. It doesn't want to learn. There might be something useful in what you try to put in, but its outside some criteria, so it's considered irrelevant. So, there may be a case where you are trying to progress a patient handover at the hospital, the patient kicks off and you have to call security, you've got a violent patient attacking the doctors, but there is nowhere to record that in the ePRF. There is no place for free text that could have been helpful about the incident.

Such discourses suggest a range of viewpoints regarding autonomy and trust. Some felt that their scope of practice and daily modes of working were excessively restricted by a blinkered and controlling management; that ambulance Trusts were bound by petty official rules and obsessed with narrow interpretations of operational performance. Others worried that the scope of practice was potentially too broad, and that paramedics were being exposed to considerable clinical and legal risk without adequate employee support. They were acutely mindful of their 'registration being on the line', and often suggested that ambulance managers did not show much care or attention to this concern. Time and resources required for training, professional development, and clinical debriefing were not being provided by the employer, due to the incessant rush to keep the roster full in the endless pursuit of response time targets. Paramedics would nearly always welcome any broadening of their scope of practice and innovations which gave them increased scope, flexibility, and versatility in the field, such as the distribution of new drugs to administer, or new clinical pathways for their patients. But they didn't want to do this in an employment context where the clinical risks, and the requirements for detailed knowledge, training, and practice, were underplayed or neglected by management. Paramedics are, therefore, in a difficult position. They have to shoulder considerable responsibility, balancing risks, and making judgements, while not feeling fully backed up by their employer.

Mad Targets

A major issue behind many of the above complaints is the role of performance measurement; the set of systems that directly and indirectly structures the behaviour of all employees in the ambulance service. Chapter 2 discussed the battery of new controls inserted into public service professions with the aim of controlling and measuring performance and providing accountability. This 'audit culture' (Power, 1997; Strathern, 2000) has reconfigured many elements of working life in dozens of sectors and professions. Although the complex nature of professional work tends to make many parts of it unamenable to management control, the 24/7 availability of ambulance services lends itself to targets-based micromanagement and audit. There are measurable outcomes associated with all aspects of the 'delivery' of ambulance services. Digital equipment on ambulance vehicles feed data into sprawling arrays of control and measurement at HQs (Beer, 2016, 2019). Data are gathered on the rapidity and accuracy of call handling and dispatching, speed of vehicle mobilization, response times, time on scene, outcomes from cardiac arrests, complaints and their resolution, roster utilization, staff sickness, and absence. The impacts of these measures on the daily working lives of ambulance professionals are potentially profound.

When I raised the issue of targets, standards, and performance measures, the usual reaction was tired resignation. Some regarded the response time targets as something largely for managers to worry about. Others would claim that the Trusts continually remind staff of their importance. I would often hear that the targets are often impossible to meet, and that roadstaff tended to put them out of their mind and focus on what they can influence. A clinical team leader, Chris, explained it to me this way, as we were waiting on station for the first callout of the morning:

> It gets to a point where the 8 minute response times are just not achievable. There is quite an emphasis on it. You've got to make the targets, and you are made very aware of it when obviously the Trust is struggling to hit them. Years

ago you don't really know the figures, but now there is a lot of focus on them. A lot of people in practice are not interested in the 8 minute and think it is less and less relevant. In a way that is right. But they *are* relevant, performance is relevant, it is no less relevant. The level we are at, it flashes up in reminders we get, as a CTL you're made pretty aware of it. Sometimes around here the traffic is so bad there is no way you can get to the call in 8 minutes, just no way. You'll see this today. The 8 minute target is kind of in the back of your mind. But around rush hour, the schools traffic, anything south of here, anything that takes you down Abbey Road, it just gets clogged up, you can't do it in 8 minutes, no chance. So it does make you wonder, how are you supposed to hit a target that is impossible to hit? How is that a good measure of performance, a fair measure of performance?

While this CTL was relatively sanguine about the targets other paramedics were more critical. In an interview with Gareth, the issue of the targets was embedded into a broader discussion of a shortage of resources, and a cynicism about the use of RRVs as a form of gaming:

Gareth: Government targets dominate our work. Targets are the only things that matter. Patient care is a side issue. Look at the FRV, or RRVs. More cars are often available than ERAs (emergency response ambulances). All the car is doing is getting to the patient to stop the target clock. Then you can be waiting a long time for patient transport, perhaps up to an hour or 90 minutes. Control will say there is simply none available. That clock has stopped, the one the car is working to. This means that further incoming calls will continue to divert the double-crewed ambulance. The target's being met, let's leave it at that. There seems to be a lot of falsification of types of incoming call. The call prioritizing is contentious. A child having a fit, a convulsion; we'd say that's a Cat A. But it may be downscaled to a Green, a thirty-minute response. And the reverse! A drunk can be categorized as Cat A. The systems are very open to interpretation. Staffing levels are a part of this. Incoming calls have been going up something like 8% year on year in the service. 12% rise at certain points. Staff levels have not increased at any level. In fact, it feels like a reduction.

Emma, a highly experienced operational paramedic and mid-level manager, had a still broader view, making some interesting comments about the targets as a marketing device, a 'ritual of verification' (Power, 1997) that is crucial in establishing and defending the 'credibility' of peak-level management organizations such as the AACE in the face of a frequently hostile news media. 'If you are below targets', she argued, 'then you look weak.' Her discussion of the risks and rewards of hitting and missing targets is reminiscent of the Stalinist logic of 'targets and terror' (Bevan and Hood, 2006), in that complaining about a lack of resources makes a local manager look ineffective in using those scarce resources:

> Emma: AACE is all about trying to look credible. You can't get more resources if you are short on targets. If you are below targets then you look weak. But you can only get more resources if you are already hitting the targets—it just doesn't make sense. Back in the summer there was this thing in the media, especially in *The Daily Mail* and *The Daily Telegraph* about what they called a 'secret plan' to change ambulance response times. It was just some memo about targets, it really wasn't anything of note. But the media will go mad if there's any suggestion that we are deliberately getting to jobs slower, or de-prioritizing jobs. Someone will die and it will be reported that they died because we got there too slow, but that's unlikely to be the reason, really. Media equates targets with care. So does management.

In practice, long distances, traffic delays, demands on the service, availability of vehicles, and the vicissitudes of call prioritizing have huge impacts on the ability of crews to hit response time targets. Just as CTL Chris had earlier predicted, while on an observation with him and his ECSW partner Sally, we ran into extremely heavy traffic that held up our vehicle for prolonged periods while on a Red 2 call:

> 0854. We get a beep on the radio. The crew quickly mobilizes and we're heading out of the station door to the vehicle parked outside. 'It's only 12 miles away' says Chris. The MDT says chest pains and DIB. We head out to the roundabout and take a left, heading into the town centre. The ambulance passes a large school and the school traffic is appalling. Cars all try to pull to the left to let us through, and oncoming traffic has to also slow and drift left. We pretty much head down the centre line where possible. Soon we get to a very large junction; three lanes in the road we are in, with several other three-lane roads connecting in a complicated, light-controlled intersection. We have about seven or eight cars in front of us before the junction begins, but all this traffic can't move as the junction is already full and cars aren't clearing it. The traffic lights cycle through two or three green cycles without anyone moving. Our blue lights are on, but the traffic is so bad that Sally turns the sirens off as there is no point deafening the other stationary drivers around us.

What a strange situation—a red call, but the vehicle is stationary for about a
minute. It feels as if the locals also realize that this ambulance can't get
through, but that there is nothing anyone can do until the traffic clears on its
own accord. Chris and Sally start to talk about which weeks they have booked
on leave. What Chris was telling me earlier about traffic was exactly right.
There is no way we'll make 8 minutes. 'Only 12 miles. . .' Well if you go at a
constant 60 mph, then that's 12 minutes. It's ridiculous.

The target times' abstractness, coupled with their often impossible-to-hit
nature, was what made some paramedics describe them as 'less and less
relevant'. What tended to be of more direct concern to responders was the
nature of the daily interaction with managers on station and with dispatchers at Control. Target times were a key background feature in this, but
the problem perceived by the ambulance crews was a broader one about
the 'frontier of control'; about workplace relations, especially around trust,
courtesy, and respect. Relationships between crews and 'Control' could be
strained. As an illustration, the below is a fieldnote taken while waiting in
the mess room for an observational shift to begin. The TV is on, tuned to
BBC Breakfast. Paramedics and ECSWs are clustered around, drinking tea
and idly half-watching the TV. Gossip is circulating:

Another para pops inside and stands by the door. He asks some of the other
crews:
'Have Control said anything to you?'
'No, nothing. Why?'
'There was a bit of a to-do. We were out, it was about 4 in the morning, heart
attack at a care home. The care worker, a young girl's been doing CPR, we took
over from her. Anyway, it went nowhere. We also needed a restock, so we
asked if we could drop the girl home, she only lived in Downingham it's pretty
much on the way back here. They said no, can you fucking believe that? Poor
girl's had the lady die on her, then we're told we have to leave her waiting for a
taxi, or walking home at 4 in the fucking morning. We had no 02 anyway, no
drugs anyway, what kind of call could we have gone to? We were heading back
this way in any case.

Disagreements like this—where 'Control' has apparently failed to consider
the perspectives and needs of frontline responders and patients—appeared
to paramedics and techs as a kind of 'micro-aggression', showing management's apparent disregard for frontline context, putting the logics of the

targets and response times above those of 'common sense'. Another classic complaint, a bugbear of ambulance crews all over the country, is the dispatching of a crew to a call some distance away when the crew is nearing the end of its shift. This guarantees that the crew would move into unplanned overtime, probably extending their shift by at least an hour, just when they were thinking about getting to the end of the shift and heading home.

There are other gripes and sources of conflict. The targets regime constructs and measures artefacts such as 'call to wheels rolling'—the time between the call being sent to the MDT and the time the crew's vehicle sets off to the call. Crews invariably mobilize quickly, but occasionally there are reasons why this target would be missed, often by a few seconds. Depending on the style and experience of the dispatchers and managers on Control, these infractions would either be ignored or would be followed up, usually by a radio call mid-shift. This kind of 'inspection' would annoy and worry the crews. I occasionally saw paramedics check, with a degree of anxiety and irritation, the PC on station to see if any 'late mobilizations' had been logged. And yet, throughout my observations, I very often heard patients and bystanders saying 'thank you for coming so quickly', even when a crew was held up in traffic or when Control had radioed in with a request for information about a 'late mobilization'.

From the paramedics' viewpoint, there will always be delays and there will always be shortages, but everyone is doing their best to respond to every call and to be caring and respectful to all patients. What they don't like is the assumption among some of the management and Control staff that some of the delays can be put down to responder laziness, disorganization or inattention. Any insinuation from Control or from station or area management that crews are taking excessively long breaks at hospital, or that their appearance or conduct is inappropriate, was very poorly received.[2]

Paramedic Dan Farnworth describes those wearing the uniform as 'stoic' and 'humble' (Farnworth, 2020: 284). That's very much what I saw. They are caring and compassionate people. They are hardworking public servants occupying a difficult role. They were often frustrated with how the service was being run and how they and their patients were being let down. Their complaints would often reflect a defence of the paramedics' own understanding of what the role is or, at least, what this role should be. Most of these frustrations struck me as justified, although sometimes the odd

---

[2] See, for example, this Tweet and the chain of replies, 19 December 2019: https://twitter.com/TomWilkes_/status/1207716818485420032

individual would come across as prickly, precious, and 'difficult to manage'. Complaining about management was extremely common, sometimes exaggerated and excessive (Hyde et al., 2016: 100-2). While this 'moaning' has been a staple feature of the ambulance culture for decades, it also reflects paramedics' exasperation with how the professionalization project is being frustrated and restricted by the NHS Trusts which employ and manage them (McCann et al., 2013).

'Working Jobs', 'Trauma Junkies', and 'the Baboon Pack': Understanding Paramedics' Cultural Values

Professionals don't seek out and defend their autonomy just to make their working lives easier. The reason they protect their autonomy so jealously is because they believe that discretion has to exist to allow them to practise their craft correctly, in our case for paramedics to be enabled to work effectively for their patients. But the way that paramedic work is organized causes considerable tensions (Newton et al., 2020). Paramedics are not in control of the calls sent their way. Many of the calls they receive are medically inappropriate for an emergency response service (Metz, 1981: 116–24; Seim, 2020: 56–61). But they have no choice but to respond to each call assigned. Paramedics want to operate in an organization that deploys them and their skills to 'genuine emergencies' or to 'working jobs' where paramedics can 'make a difference'. But, time and again, they find themselves not attending 'good' or 'working' jobs. Instead, they are sent to 'dross', 'crap', or 'rubbish' (see also Chapter 6).

This is an interesting tension in ambulance work. It was common to hear roadstaff describe the importance of their work, and the 'privilege' of responding to patients in desperately difficult emergency scenes. They resented their time being wasted on medically unnecessary callouts. They wanted their clinical skills to be used appropriately, and nothing compares to the 'high' that comes at the rare occasion when a paramedic crew has genuinely saved a life, such as at a major trauma call, or a cardiac arrest. They also spoke glowingly of their clinical practice and pathways being expanded to allow them to contribute to 'Gold Standard' interventions such as PPCI for STEMI emergencies.[3] One paramedic described the experience of watching a STEMI patient recover as 'miraculous'. To a large extent, they

---

[3] PPCI (Primary Percutaneous Coronary Intervention), also known as angioplasty, is a procedure for treating narrowed arteries. The Gold Standard for out-of-hospital treatment for a particular type of

actively seek out the most clinically serious calls, as these calls are seen to make the most efficient use of the scarce paramedic resource and represent the summit of the profession's meaning. Ambulance culture features an informal status hierarchy of calls in which life-threatening conditions such as major trauma, strokes, and heart attacks are the most important, interesting, and valuable callouts that provide a test of a responder's ability and opportunities for clinical learning, whereas an array of less clinically serious calls such as minor injuries, elderly falls, anxiety attacks, and children with viruses are seen as clinically dull, a frustrating waste of scarce time, and totally forgettable. Life-threatening emergencies often have an attractive simplicity, making them different from the typically ambiguous and inconclusive 'social-care' type calls. There is normally a clear and unambiguous series of tasks for the crew to complete as they try to prevent death or treat catastrophic injury. These calls are intense and highly focused, where paramedic training 'kicks in'. Paramedics don't shrink from the risk that critical cases provide; instead they embrace it as the most undeniably important part of their profession and their identity (see Chapter 6).

It is this attraction to the risk, danger, and social importance of serious emergencies that has led some of the literature on ambulance work to explore a notion of paramedics as 'trauma junkies'. This term features strongly in the work of Eddie Palmer (1983). Tangherlini's book and article (1998, 2000) about paramedics' storytelling is similar. These are important contributions to the literature on US paramedics and EMS, in which responders are portrayed as revelling in extreme stories and the 'trauma junkie' persona. But, in my study, and in a similar way to Seim (2020: 51), I found the 'trauma junkie' idea to be somewhat inaccurate, for two main reasons. Firstly, we have already seen how ambulance services in England are substantially moving away from the explicit focus on life-saving emergencies. Secondly, the 'junkie' persona has certain 'old school' connotations that do not really square with the increasingly ascendant notion of the paramedic as a clinical professional.

The trauma junkie notion did partially appear. Several paramedics shared with me some outlandish stories, some of which were as colourful as those in Tangherlini (1998, 2000). A small number of paramedics I interviewed and observed exhibited at least some of the trauma junkie identity, but a more NHS-type, clinical care persona was generally much more common. On several occasions I was told that the paramedic role is fundamentally (and legally) a clinical NHS role, not an emergency

heart attack known as STEMI (ST-segment Elevated Myocardial Infarction) is for a paramedic crew to take a patient direct to a specialist PPCI lab rather than the general A&E unit.

services, uniformed role. And, while paramedics put very high value on the importance of attending 'genuine emergencies', they would not want a shift featuring an uninterrupted string of high-pressure, high-risk, stressful, emotionally challenging, and grisly callouts. There were times when paramedics clearly valued quieter shifts where the calls are low risk and not life threatening. Providing care to patients and acting with understanding and compassion are all valued activities that are taken seriously. Paramedics are professional care workers, not thrill-seeking adventurers.

And yet, within that broader care identity, the 'edgework' element of emergency response was a major part of the attraction of the job (Granter et al., 2019). Ambulance work has a unique nature, quite different from many other NHS occupations. Paramedics' clinical, NHS persona is strongly imbued with certain cultural features associated with uniformed emergency services. Ambulance roadstaff enjoy the immediacy and autonomy of the work. They like not knowing what is coming next. They greatly value the opportunities that the work provides to interact with patients, bystanders, and other medical professionals in emerging, unfolding, and unpredictable environments. Ambulance work can be frustrating, unforgiving, and stressful, but it is rarely boring.

I observed a shift with Katie, a Paramedic Practitioner, with a nursing background. She was formerly a nurse on a High Dependency Unit, an experience she found depressing and uninvolving. Compared to HDU nursing, working on an ambulance provided more interaction, much more variety, and often more hope for better patient outcomes. She also enjoyed the sense of adventure that mobile, emergency response work provides, as opposed to being stuck in a ward for hours on end:

> I've been in the ambulance service about 14 years. I used to be a nurse. I went to Crayton Polytechnic. I was working at [Hospital] on HDU and I basically hated it. The patients were so ill, so infirm, you often couldn't talk to them, it was hard to get to know them. They would die frequently, there was often nothing you could really do for them. I think maybe looking back with hindsight I got quite depressed working there. The reason I left was because we got to do a shift third manning with the ambulance service and I really enjoyed the experience. It was totally different from being on the ward day in, day out. It was much more varied, more interesting, a lot more relaxed in a way. You also have more autonomy out here. As a nurse you were always asking the doctor: 'I'll go and ask the doctor', 'ask the doctor' . . . you had no autonomy. You get brainwashed. But in the ambulance service you are the only one there and you have to be in charge. It's down to you. There is no-one else around.

In addition to the autonomy and freedom, street work also provides ambulance staff with access to strange, noteworthy and 'extreme' encounters (Granter et al., 2015; Metz, 1981; Tangherlini, 2000), often involving physical and mechanical destruction. In the following interview, David describes a surreal callout involving wrecked vehicles, armed police, and references to action movies such as *Mad Max*:

David: [It was] the first time that something came up on MDT as 'extreme fall'. It was a really unusual job, it was in Drainford and we at first couldn't find the place on satnav, and had to buzz up to ask Control where it was. We were directed towards a bridge over the road and, as we got closer, we could see lots of police – armed police – and a vehicle that was in a really bad shape, broken glass everywhere, lots of damage to the car. The car was up on the bridge but the patient was down below on the street with police standing around him. He was in bad shape, writhing around on the floor, was in severe pain and had injuries that looked very serious. It turns out it was a police pursuit. They'd crashed the car and one of occupants had panicked and jumped from the bridge. It was quite a drop. Either he didn't look or just was in a panic, I don't know. The scene was very dramatic, the old disused factories and warehouses everywhere, the scene basically deserted except for us and the police and the patient. There were sounds from some of the buildings, industrial sounds, machines banging and humming. It was like *Mad Max* or something, a movie director couldn't have found a better location!

Anyway, he was very lucky. When I returned to that hospital later in the shift I asked how he was. Turns out no apparent major trauma, it looks like he'll be fine. But to jump from there—just crazy. My partner said he was one step away from an entry into the Darwin Awards!

Calls involving adventurous and dramatic settings often provide good material for storytelling and sometimes synchronized well with paramedics' non-work interests and general 'action-oriented' outlook on life. Even as their job was physically demanding and sometimes dangerous, I noticed that many of the paramedics I interviewed or observed had interests outside the job in outdoor 'edgework' type pursuits, including extreme sports. Hobbies of paramedics included rock climbing, mountaineering, sailing, kayaking, performance motorcycles, triathlon, running, and outdoor swimming.

Palmer's (1983) notion of the trauma junkie clearly has relevance. But times have moved on. The operator culture is considerably more moderated now than from the time described by Tangherlini (1998, 2000), especially as seen in videos he made that accompanied his work. Some of these clips survive on YouTube.[4] Watched today, many of Tangherlini's EMTs and paramedics appear as hyper male, borderline racist adrenaline junkies. They laugh about, boast, and re-enact scenes involving automobile accidents, gunshot wounds, and attempted suicides. While some of them are talented raconteurs, they come across as ambulance cowboys. Echoes of this culture still exist. The paramedics I met were sometimes candid in their discussions with me about bizarre calls in private, but it would be unimaginable to point a video camera at them and encourage them to retell stories in such a fashion. Speaking publicly like this would expose them to likely suspension from their NHS Trust and possible investigation and removal from the HCPC register on the grounds of 'fitness to practise'.

'Professionalism' has a lot to do with informal and intangible expectations of behaviour and conduct. While increasingly careful about external appearances, paramedics today continue to exchange weird, tragicomic, and grisly stories among themselves in private. Dealing with the strange and unique circumstances of serious emergencies remain a central indicator of the informal value and meaning structures that ambulance staff attach to their work. Many of them are drawn to roles that promise a greater proportion of 'genuinely life-threatening' calls. Being seconded to an air ambulance post was an ambition for many. Critical Care Paramedics do not exist across the country but, where they do, they tend to attract paramedics willing to take on a trauma-focused ethos and identity. Others would comment on the style and practice of CCPs, often suggesting that CCPs are the types who embrace maximum autonomy and are confident with clinical risk. A comment made while on the road with solo paramedic Emily was interesting:

> Yeah, you're often struck by how different they are, how bolshy they are. They live for trauma, CPR jobs and never seem fazed by it, just a stream of those kind of life-threatening jobs. Other paramedics see the CCP arrive and they think 'phew'. That's a new thing, really, you never had that level of reassurance before.

[4] Some video clips of Tangherlini's documentary are posted here: https://www.youtube.com/watch?v=fZOWswl83bQ
https://www.youtube.com/watch?v=ng4D0g3iY7c&t=165s

Observing a shift with a CCP crew provided some insight into their desire to 'live for trauma'. The crew started their shift at a major ambulance station, but would then relocate to operate from a 'tethered' station which, as part of the Trust's deployment plan, was near several major roads to maximize availability. As expected, the crew regularly went to very serious calls involving threats to life, and was exempt from responding to Category C calls. But the vagaries of patient demand and constant movement of vehicle location meant that they would also sometimes find themselves tasked to much less serious calls. It was fascinating to see how they had developed an informal practice of attempting to 'hang back' from 'jobs' that clearly aren't the major trauma, strokes, or cardiac arrest calls that they would hope to be assigned to:

> After about twenty minutes' drive we seem to be manoeuvring to park somewhere. Becca jumps out of the passenger door. I hear bleeps as we reverse slowly into a narrow garage doorway. Pete calls out, 'ok Leo, we're parking here for a bit. You'll need to jump out the rear doors, as there's no space for the side door, mate.'
> I open the rear door and take in my new surroundings. It's a pretty small, well-lit garage area. It is very messy. I hop down from the ambo and look around. The garage is nearly full with junk of various kinds. There are filing cabinets, boxes of old paperwork, planks of sheet metal, retired old ambulance station signage, a few doors into other little rooms. The ambo itself takes up about a third of the space. What is this place? 'This is Riverside. It's what we call a 'tethered' station. We're near to motorways and main A roads.' Pete ushers me through a doorway into a little mess room, smaller than I'd seen at other stations. There is a little kitchen area and a table, and a large TV and a few comfy seats at the other end of the room. It's a very small ambulance station, big enough for one or maybe two crews, really.
> Pete offers to make tea, Becca is constantly tapping and swiping on her phone, talking about holidays, and seems to have messages going back and forth between her and her boyfriend about where they plan to go. Nothing much is happening, but that doesn't last long. There is some activity on the radios, and we're heading back to the vehicle. They move with a quickness but not a rush. There is just about enough time to pour the cup of tea down the sink and bin the cups. I do need to kind of hurry up though, and I open the rear doors and climb in just as they get in the front. I do up the seatbelt and we're off. I check the time—0728. We'd been at Riverside less than ten minutes. Becca looks at the data terminal: 'Female, unknown age, fallen. Blurred vision.' I don't recognise the location, but they tell me it is only 3 or 4 miles away. With blue

lights on and not much traffic out here at this time of the morning, plus the occasional blare of the siren, we get there very quickly. Further details of the call feed through onto the computer system enroute. 45 years old. 'What's a 45 year old doing falling over?!' says Becca. More updates through the system: the patient has an underlying health problem. It's starting to sound like an unplanned primary care type of call.
The crew start to talk about how to deal with this, and whether or not it will get downscaled to a Category C call. Soon exactly that happens, 'Cat C!' says Becca. She tells Pete to 'Hang back, wait here.' She explains to me through the window into the cab: 'Sometimes you can hang back, but if the patients or bystanders see you, then you've then got to commit to the job. But sometimes it's a good idea to just hang back if someone else should be going to this kind of job.' Looking at the sat nav, we are literally around the corner from the patient's house. But her plan doesn't work. The radio sparks to life, 'Would you be able to have a look for us?', asks Control. Becca: 'Ok, received, that's absolutely fine, we'll go and see.' The attempt to leave the call to another crew hasn't worked. Obviously control can see that we are literally twenty seconds from the job. Becca's response on the radio was extremely polite and friendly, and just ever so slightly overdone, which sounds as if she is including just a suggestion of sarcasm.

A large part of 'living for trauma' stemmed from the concern shared by all crews that ambulance trusts need to be sufficiently well organized to allow sensible deployment of resources so that patients most in need received the most appropriate response in a timely fashion. But there was another element to this, too. Those who regularly handled trauma calls were usually respected and admired. And for any paramedic—not just the highest trained CCP, PPs, air ambulance paramedics or members of the specialist HART or SORT teams—experience gained from high acuity calls were a kind of reputational commodity traded on station. When station gossip would start to include fragmentary information about the conduct of an unknown or junior paramedic at a major emergency, this meant that this person was starting to become accepted into the ambulance culture; that strange hybrid of uniformed emergency service and NHS clinician. Below is Simon's take on this:

Simon: Have you read any of the military literature? Memoirs and stuff?
Leo: Yes, lots of it.
Simon: You know that saying 'FNG'? The 'Fucking New Guy'? Now it's not like the army in a warzone where you don't want to get to know a new person as they

might get killed and you don't want the pain and sorrow. It's also different from Fire, in that you're not thinking 'is this guy strong enough, is he going to be able to pull me out of a fire?' So it's not about danger or of covering yourself. But there is an element of the FNG thing in that you don't know what they are like. When a new guy arrives, what are they going to be like to work with? When I look up the name on the orders, I think, 'who is this new guy like, what will the shift be like?' That's all I want for me. I want people to think of me that way. I want people to look on the roster and say 'ah, it's Simon today. This will be fine, this will be a good shift.'

It is a bit intimidating at first when you are a new guy. I've been the new guy. You can't just join in with the banter and the gossip. You don't know what people think of you and you're not sure you can become accepted. You look in the station, and the guys are sat there on the comfy seats talking and joking among themselves, I call them the baboon pack. But for whatever reason people will hear about you, you'll have something to contribute, and one of the baboon pack will eventually invite you in.

Simon's notion of 'the baboon pack' is a telling metaphor for one aspect of paramedic identity. In this case, the baboons sound very much like a group of 'old school', male paramedics, whose approach reinforces traditional notions of toughness, reliability, and credibility. A person would only be invited into the 'pack' once he or she has 'proven' themselves on the street and had their status confirmed through the station gossip system. The focus is often on how novice paramedic handled themselves on specific 'jobs': 'I heard you attended that motorway RTC the other day?' 'Were you at that suspected murder job?' Station gossip can focus explicitly on how specific incidents were handled, but can also function in more general, more subtle tones, as staff gradually figure out who among them are the most effective clinicians, the most supportive colleagues; who is or is not the best person to be crewed with.

The informal actions and judgements of station gossip are an important part of ambulance service culture. It has powerful implications for how notions of a competent paramedic are constructed by crews, and how the 'professional' role of the paramedic is often contrasted against the restrictive and unwelcome behaviour of 'management'. Roles and behaviours deemed appropriate to paramedic culture are endlessly recycled and reaffirmed through station gossip, especially through joking, griping, and banter. The next section explores this element of the ambulance operator culture in more depth.

'We Can Moan for England': Decompressing, Joking, and Complaining

I opened this chapter with a quotation from a classic piece of sociological writing on the social roles of uniforms (Joseph and Alex, 1972). To my eye, it seemed that the paramedics took pride in the uniform and their appearance was always highly presentable. But I found that the uniform itself did not have as much of the 'behaviour formatting' effect that I was expecting. Paramedics spoke in an informal way to pretty much everyone they interacted with, especially to patients, but also to Control on the radio. They had an NHS identity, much like nurses. They would interact in a similar way: relaxed, personable, usually with exceptionally good interpersonal skills. Although the uniform has a somewhat robust look to it, and ambulance stations can be somewhat intimidating places, those wearing the uniform were not at all like the caricature of a brainwashed soldier or the tough bearing and dominant presence of a police officer (Coser, 1974). They wore the uniform lightly.

They frequently work solo, operating in quite lonely environments. They have considerable scope to figure out their own working style. I illustrate this with a fieldnote from a shift with Paramedic Practitioner Katie. It suggests some of the loneliness of the solo responder, here juxtaposed with a description of a classic piece of inter-service rivalry with Fire and Rescue:

> We have arrived back at that standby post in the near-empty fire station. Katie mentions how quiet it is and how they hardly have any interaction with the fire crews. In fact, she even mentions sometimes the fire crews complain about noise from ambulance staff on shift! 'If you're on a night shift you can hear them snoring!' She points out the military-style drill yard around the back, 'they sometimes line up here and have uniform inspections. The way of working for fire and ambulance are very different.'
> I head down a corridor towards the toilets. Dusty winter sun is slanting in through the windows. It catches a row of yellow-brown fire service uniforms hanging from a rack. It is indeed quite a sleepy scene. A fire service poster on the wall instructs staff to 'not mobilise the Water Rescue Unit for any reason at night time hours.' Another warns about security, saying something about 'we have recently had intruders'. I notice an old sticker on the inside of the toilet door advertising a Fire Brigades Union industrial action demanding £30k/year.

Someone had scrawled their own comments underneath: 'NOT NEEDED', and:
'what about paramedics, techs?'
I head back to the little NHS staff room again. I mention the union posters.
Katie: 'Most of us here are Unite, not Unison. Part of this area was known for
union militancy at one time. We were known as the People's Republic of North
Disfield. People do like a moan.'
The radio bleeps again—54 year old female with chest pains.
'It's very nearby, we'll make 8 minutes hopefully!'

'We do like a moan' was an extremely common refrain. Incessant complaining and whining was deeply entrenched. Moaning would almost never
take in front of patients and bystanders, but I witnessed a great deal of
it in backstage areas, as well as a lot of comment about the overall phenomenon of ambulance service moaning. Indeed 'the moaning' was such
a well-recognized element of ambulance culture that some described it entirely as a cultural artefact, part of the 'old school' style of ambulance work
and, interestingly in the following account, 'not really related to anything
real':

> Sarah: That old school culture – that is starting to disappear. It is less common
> now, but you see it a lot in the middle management ranks. Some of the moaning
> and whingeing is a cultural thing, it's not really related to anything real. And you
> see this with the new recruits out of university, they are from a different back-
> ground to the old school roadstaff but they immediately take on the bitching and
> moaning, the suffering. But you look at it and so often the shift is routine. Yeah, it
> may be a long shift, 12 hours, it is tiring, but often the work is steady, there's been
> nothing life-threatening, no cases where you've been forced to rush, say you've
> had 7 jobs today, nothing urgent, nothing taxing. A slow handover at hospital
> which gives you a bit of a rest, you cleared off the job and didn't get something
> straight away, you know, a steady, plodding day. You do get those. Days when
> you don't remember any of the jobs. But is not in our nature to recognize that.
> Instead its moan, whinge. . . It's hard to get people into a place where they can
> see the positives of the job. [. . .] The new recruits, they follow what they see,
> and they pick that all up quickly.

Rob, a highly experienced and clinically very well-respected paramedic,
spoke on this matter in detail. Although acknowledging the low morale, he
was keen to emphasize that moaning all the time is pointless, unpleasant,

and self-defeating. He described it as 'a stuck record', 'wasting energy', and 'dragging everyone else down'.

> Rob: The moaning is interesting. On some level it is totally accurate, you know, these are genuine problems, the terrible management, the call prioritizing, the general lack of care and attention. Then there's the patients who call for stupid reasons, and all that element. But the constant moaning is a problem, it is cultural. It is like a stuck record. Sometimes, yes, you are exhausted, it is hard, it is bloody hard, but the constant whingeing and moaning just makes it worse. You are wasting energy, dragging everyone else down, too. Sometimes I join in, other times I'm like, 'right that's enough, this isn't helping you or anyone else'.

Other terms I came across for 'the moaning' was 'pissing and whining', 'venting', 'huffing and blowing', and 'standing around with hands in the pockets'. It was interesting as it was something everyone admitted to, but was increasingly portrayed as something associated with the 'old school' culture, and not something that 'professionals' should engage in. It was also very instructive to see this attitude ascribed to experienced 'old school' paramedics who have since moved up into management roles. The circulation of this perspective further served to discredit management as out of touch, burnt out, demotivated, unsupportive, and just generally unpleasant.

'The moaning' also had a peculiarly self-depreciating angle (see also Cosineau, 2016). It portrayed a quite different feel for the ambulance service from the 'hero' narrative that tends to feature heavily in TV documentaries and the 'I could never do what you do' line so often heard from the general public (Jones, 2020: 6). Rather, a 'moaning' ambulance worker will typically describe the ambulance service in terms similar to those used to describe a rundown industrial company with its best years long behind it. They describe the service as underfunded, underequipped, and poorly led, struggling with barely functioning vehicles and inadequate training. They will describe much of the callout activity as easy, futile, and not really worthy of comment. Similar ideas have long been expressed in blogs such as *Broken Paramedic*, or *Big Yellow Taxi*.[5] One paramedic, David, suggested that the constant complaints on station are becoming 'more like a joke, a parody of the moaning'. The social media accounts of some ambulance

---

[5] https://westhorpe.net/category/broken-paramedic/ https://adventuresoflondonsbigyellowtaxi.wordpress.com/

staff would often contain similar content, such as the timelines of 'Stroppy Ambo Woman' and 'Toytown EMS'.

'Moans' could sometimes have a political inflection. As in all other parts of the NHS, ambulance staff were committed to the concept of the NHS as free at the point of use. They would air complaints about the creeping privatization of the service, such as the growing dependence on private crews and volunteer community first responders. Some of the more experienced paramedics also shared in-depth knowledge of local level NHS politics, commenting on the availability of services, which hospitals are heavily in debt due to PFI deals, and the highly suspicious retitling of ambulance 'localities' as 'business units'. Complaints about specific political elements of the workplace tended to be somewhat less common among the younger staff.

The following fieldnote illuminates many of these elements hanging together in everyday ambulance service station discourse. I am accompanying PP Katie, as we head to a standby post for a meal break. Two other crews were also on their break when we arrived. Discussion among the three crews very quickly turned to management and employment issues. Everything is bound together—stress, inadequacies in the system, poor management, lack of public understanding and recognition of what the paramedic job entails, guffaws at media coverage of the service, complaints about others seen as clinically unsound, and the use of private contractors. Even here we get different views, from a dyed-in-the-wool public sector unionist to a younger paramedic who seems uninterested in union talk of disciplinaries and job evaluations and refuses to get drawn into 'the moaning':

> The standby post is a ground floor office with reserved parking spaces outside. It's a council building, but the usual ambulance station stuff is all in the room allocated to responders: big TV, easy chairs, a computer in the corner. Katie goes to check her email: 'let's see if I've got any complaints,' she jokes. There is another DMA crew there, of two men, one maybe in his thirties, the other more experienced and wearing a Unison tabard, as well as a young man working alone on the SRV. I listen to them talk for a while about various work issues, pay, holidays, managers, training. They mention that TV show *Emergency Bikers* and have a laugh at one of the 'Brummie' paramedics on the motorbike. The young guy mentioned a clip from the show was used in training where he'd gone to a CPR job but didn't bring his defib as he walked into the house. I've noticed this sort of thing quite a lot across the ambulance world—people criticizing others for supposedly not being up to the job.

The older guy then reflects on the issue of public appearance. 'Management is very concerned about bad publicity. If you complain or moan about any element of ambulance provision in public and your name comes out you are effectively signing your own P45. The rule is really not to talk to the media at all.'

He's the Unison rep and seems a really nice guy.

When I mention the paramedic role growing and getting more sophisticated they get into a long conversation about banding and pay. It becomes a classic ambulance station moan-fest.

Young guy: 'I don't get how we are band 5 when other occupations are higher. I mean, art therapists, for fuck's sake.'

Union rep: 'The job evaluations didn't make much sense and the different roles in ambulance trusts weren't understood or properly evaluated. Techs should be band 5 but became band 4. Paramedics were not far off band 6. They were about 8 points shy of the threshold; we came very close. They took away our intubation skills which was a major part of it. In hospital roles the kind of work that PPs do would be much better paid, on a much higher banding.'

Talk gets on to the huge stress levels in the service, including among managers. A lot of people are on sick leave and don't want to come back. Some are checking their own blood pressure, getting doctor's notes and the like. There's conflict about whether people are really well enough to work. This is also 'driving managers mad' as they can't fill their rosters.

The other guy working solo and watching the TV was quieter and in truth less friendly. He said something about all workplaces being tough these days and no-one is having a good time. There is no use whingeing; you have to get on with it. 'Adapt and survive', he says. His words and appearance suggest to me that he's ex-Army.

The composition of the ambulance workforce is changing. Moaning was not the exclusive preserve of the disaffected old school responders. As Sarah suggested, a cynical outlook and a culture of complaining rapidly passes onto the newcomers, including the degree-trained paramedics. We will see in Chapter 6 how the promise of professionalization is slowly taking the paramedic occupation on a more ambitious upward trajectory, but how it cannot generate an escape velocity powerful enough to break through the gravitational pull of the troublesome climate of ambulance work; procedural controls, questionable call prioritizing, and an overwhelming emphasis on processing volume of calls rather than developing clinical skills. The phenomenon of 'the moaning' is a perfect reflection of this frustration.

Gender and Diversity

We have seen that the ambulance service culture features a range of unpleasant problems, such as roadstaff-management conflict, poor morale and high degrees of burnout and sickness absence. With so many problems circulating, it was interesting to learn that gender relations were generally not seen to be one of them. Rachel, an experienced IHCD-trained paramedic suggested that gender relations is 'not something [she has] thought about much':

> Rachel: It was male dominated when I started. Now I feel it's gone the other way, it seems more female dominated. A lot of the middle managers are women. It's not something I've thought about much, it doesn't bother me. There is that element that makes it difficult in that this is a physical job. There are some situations where women are usually smaller and there is that need for more physical strength that you'd tend to get more from a man. Sometimes I feel more confident working with a man for that physical support, possible extra protection if you are working nights, I'm maybe feeling a bit more comfortable if I'm crewed with a man on night shifts. There's not much banter these days. It's become much more PC. Working with men, with women, it's not really an issue.

Simon had similar views, suggesting there is no 'problem' or 'issue' of gender relations in the service. Characterizing female ambulance workers as 'quite intimidating', he used the notion of 'acceptance', in much the same way that any newcomer on station has to 'establish yourself'. Interestingly, any possibly expectations of patients around gender were also similarly described as 'not an issue':

> Simon: I really don't think it's a problem. Women are accepted on station. Maybe they behave a bit differently in general, but they are respected if you've established yourself. For patients the boy and girl thing is not an issue. I've never had a situation where a woman has acted in a way in which she didn't want me there and wanted a female paramedic. It is an emergency and this is not really relevant. They are relieved to have the ambulance arrive.
>
> People sometimes think it will be an issue when it's not. I was crewing with a woman ECA and we got a call to a mosque—she starts saying 'oh, it will be an issue, they won't want me in there.' But I said, 'Well, let's wait and see', and, sure enough, it was no problem, they were pleased to see us, and no-one mentioned a thing or acted in a way that showed discomfort with either of us. Only on a few

occasions have I been aware of being a man and this being awkward for me or the patient. I've been part of a double male crew at a women's psychiatric ward, for example.

I think the relations are pretty much equal. In a way it is easier for single women in this occupation than in others, they are very unlikely to get hit on by men at work. There will be joking and banter but a paramedic woman is quite intimidating in a way and men aren't going to try to belittle them or chat them up or whatever. It doesn't really happen in my experience.

As he elaborated on this subject, a slightly different element emerged, particularly as he started to discuss one specific colleague. Rather than notions of toughness, equality, and 'acceptance', here we see a kind of private intimacy associated with working together closely on the road. Male and female ambulance roadstaff work together in enclosed spaces, in an 'emotion-laden context' (Boyle and Healey, 2003) where persons are exposed to the sadness, shock, grief, and anger that are often the territory of emergencies, death, and injury. Under these conditions, certain colleagues would form quite strong emotional bonds:

Simon: Lucy was my student when I was a mentor. Before I worked in the ambulance service I'd never really worked closely with women before. We trust each other implicitly, we don't have to even discuss stuff, we know how each other works. I crew with her from time to time, or I'll see her out on the road when we're both at a scene. We both know we can say anything to each other, discuss absolutely anything. The physical proximity, it is close, it is a close working relationship. With Lucy we studied at the same time and became paramedics at the same time. So we've both grown into the role at the same speed, have similar experiences and interpretations. She's cried on my shoulder, I've cried on hers. I've seen things with her and gone through things with her that my wife won't know and would never experience with me. My wife doesn't know me the way that Lucy knows me. Her husband won't know the same woman I know. It's a closed space, you are sometimes touching as you move around, you know, you may touch your partner's body as you move around behind them. I can remember once accidentally touching her bra strap under the shirt, and you sort of think 'oh, shit!' but there is no reaction from her at all, it is just the way it is, we both understand. One time she was placing the ECG dots on the patient's body and I'm moving around to get the BP cuffs and then I have to kind of loop the wires around behind her, and I end up kneeling down with my head practically

on her knees, but it is nothing at the time, and you are both focused on the patient and it would be wrong to say some silly joke or to make anything of it. It's
the same with a man. Let's say you have both hands occupied with the patient
and you need your thermometer which is in a shirt pocket—your partner knows
this and he just reaches in to get it out. It might seem awkward and odd, but it
is normal, you don't make anything of it.

I'm not fully convinced by his line that 'it's the same with a man', but I appreciate the sentiment Simon is conveying. There is definitely an aspect of the
nature of emergency work that transcends gender dynamics in that everyone who is part of this unique culture understands its demands and norms.
When trusting relationships are established, colleagues are able to 'discuss
absolutely anything' with each other to provide informal emotional and
professional support (see Chapter 5). Simon's viewpoint hints at the existence of unspoken norms about the shared experience of the profession,
not unlike the connotations surrounding 'wearing the uniform', suggested
by Joseph and Alex (1972). But not all paramedics and ambulance staff will
share such close relations, as Simon himself showed with his metaphor of
'the baboon pack'. Paramedics affected by distressing calls will not seek out
emotional support from persons they don't trust. Instead, they are likely to
suppress their emotions and 'bury' the trauma, for fear of looking 'weak';
an issue with clear gendered implications given the legacy of the ambulance
service's 'macho' uniformed culture (Mildenhall, 2021: 60–1).

This is clearly still a problem, although female paramedics would usually note that this aspect has improved significantly from the recent past,
noting that former eras had different uniforms for women and men, and
where sexist and exclusionary practices were common on station. Overt
gender discrimination and sexist behaviour from other colleagues was not
generally reported to me as a significant problem, although allegations of
sexual misconduct sometimes appear as part of broader complaints into
'toxic' cultures in ambulance trusts.[6] Several women noted insulting comments from male bystanders, often about 'women drivers' (see also Walder,
2020: 204). Attending a College of Paramedics meeting, a very senior male
paramedic described to me what he saw as profound change from the sexist
assumptions around perceived difficulties of 'incorporating' women into
the ambulance service that circulated in the 1980s:

---

[6] 'East of England Ambulance ex-staff speak of 'toxic environment', BBC News, 11 October 2020.

> Management would use ridiculous excuses: 'it's not the women, its more that men can't be trusted on station with female co-workers especially on night shift.' Or that adapting stations to provide female toilets and changing areas 'would be cost-prohibitive'. All the usual, totally discredited excuses.

While gender relations may have improved, ethnic diversity remains a very problematic area. As a researcher, it was a difficult theme to get a handle on. The ambulance workforce is overwhelmingly white and British, with just over three per cent BAME representation.[7] Almost everyone I met during the observations, and everyone I interviewed, was white and British. In addition, and unlike other professions in the NHS, the numbers of migrant workers employed as ambulance staff were low. I came across the occasional paramedic from mainland Europe and from Australia, especially in the Trusts that covered large cities. David was one of the few paramedics to mention this:

> David: The old guard has mostly gone, the old school paramedics, they are now replaced by young Aussies; they are more pliable. They don't mind doing the relief rota. I can't remember the last time I crewed with any of the real old school paramedics.

The steady arrival of paramedics from overseas seemed to be an outgrowth of national shortages of qualified paramedics. A highly experienced paramedic and locality manager, Carl, explained this to me in some detail. As with many other ambulance personnel I interviewed, he positioned this discussion within a context of a slow-moving and mistrustful managerial climate:

> Carl: It used to be a ratio of about 65% paras and 35% EMTs. Now the ratio is reversed, with around 65% being techs. We are currently 50 paramedics short, just in this city. We can't get people to work in the city. New students are desperate to work somewhere else. You have 10-12 jobs per shift and it is very demanding. Many also do this on top of a long commute. It's too much. One thing I worked on was to try to find new staff. I looked into this for a few weeks and thought about recent migrants to UK. I noticed there was a large Polish community here, for example, right? And I thought, 'what about paramedics from

---

[7] Ambulance Trusts and the representative bodies struggle to address this acute imbalance. In June 2020, the College of Paramedics announced its 'Commitment to Change', in which the Chief Executive noted that it is 'unacceptable to ignore this'. https://www.collegeofparamedics.co.uk/COP/News/Commitment_to_change.aspx

Poland? How are they certified and trained?' I looked into it and these guys were good, they were trained to a high level, and would likely come here for the money we pay compared to back home. They had a first degree and some a Masters-level qualification. It was hard to get this kind of information. [. . .] In the end I had to put in a Freedom of Information Request, would you believe? But I did it and, yes, it worked. You have to be extremely specific in what you ask for or they reject it. So I asked about HCPC registration of paramedics from various east European countries over the last three years and they gave me back a very detailed response. It looked to me that this was viable, they are a growing community in the NHS and then you've got the language issue as well, why not go for this? I recommended to management that they look into an overseas recruitment drive. Typically, nothing was done with my report. But then about 18 months later I had a little 'knock, knock' on my door. 'Er, Carl, er, about those Polish paramedics. . ..!' [Laughter] It's ridiculous, I know. But anyway, at least they eventually looked into this and we have been recruiting overseas to fill the gaps. We've got several from Poland here in nearby stations: 8 or 9 in this locality, plus others, a couple of French, a couple of Aussies, a few Canadians.

The glaring lack of ethnic diversity in the ambulance service is increasingly an issue that senior leaders are attempting to address, and not without certain tensions. I interviewed a very senior figure at one ambulance trust, who argued that long-standing issues around professional standards, supportive line management, and gender and diversity issues are all intertwined. She argued that many elements of the paramedic working experience could improve if there were root-and-branch changes to roadstaff operator culture. It needs to become something 'more professional' in a sense that both management and clinicians could recognize and support. Her account hints at conflict over what is considered 'professional' and who gets to choose. Her account includes some interesting points about the limitations of the 'emergency' culture, about how old-fashioned, exclusionary, and condescending its culture can be towards outsiders, especially towards female outsiders. From her perspective as a senior leader, the culture needs to change towards a 'cleaner', more transparent and more diverse form of professionalism that transcends the 'old school' culture. But the dyed-in-the-wool ambulance people typically want to defend the sanctity of the ambulance station canteen culture, arguing that there are hidden advantages to its robust and uncompromising nature that management refuses to acknowledge.

Linda: The NHS is majority female, but the ambulance service has a long history of being male dominated. It has changed rapidly at operational levels, and paramedic students are now predominantly female. But at middle management level and above it remains very male dominated, even today. It was funny, not long after I joined I mentioned needing to go to a station, and one of the male middle managers suggested that he come along to 'escort' me—that was just hilarious to me. Anyway, when you get on station you get all that 'we never see management', 'management knows nothing about what we do', etc. There was one station that had a big pencil drawing framed on the wall, a caricature of a lot of the crews who worked out of that station. There was obviously a talented artist there who did this sort of thing in his spare time! On one level it was quite nice, it was comical, it showed they were a close-knit team, but there was one woman on there who was depicted in this ridiculous, sexualized way, with these huge boobs. I was pretty horrified by that. I said to a female paramedic 'goodness, look at this', and she said 'yeah, isn't it great?!' I'm not sure if it was meant to be her, or someone else! But in any case she liked it. . . I don't know. . . [laughing]. I mean, the culture on station, it was like a mixture of *On the Buses*, *Love Thy Neighbour*, and Benny Hill.

[T]hey'd be these persistent behaviours, typically written off as 'banter', 'just having a laugh'. For example, I was once with [Steve] and another area manager came up and said 'ah, Steve is a poof, but he's alright, he's our kind of poof and we love him', something like that, it's just so old school and embarrassing. But the LGBTQ network is thriving now. It something I'm very proud of. Ambulance services have a poor record of diversity among the NHS. But we have made tremendous progress.

Leo: It's funny, there is this holdover of old-school cultures, but many paramedics I know are actually very liberal, very relaxed, partly because of what they do, they see literally everything, which makes them very philosophical, very non-judgmental. Mostly.

Linda: Yes, that's true, they are normally very relaxed, and very open to diversity, LGBTQ, they are supportive. This is why I think there has been a lot of progress, no-one really disagrees with the idea of equality and diversity. I remember having a long discussion with a paramedic who is BME, and she was very upset about an incident when an elderly patient used the 'N' word. It was a nasty incident, it's in no way acceptable, but are we really going to make a formal complaint against a woman in her nineties, you know? Anyway, I discussed this with her a lot and we got onto the diversity of the [Trust] workforce and it was very constructive. She said something interesting, in that yes, the racialized, sexist, anti-gay so-called 'banter' has to go, but she said it's important we don't

also lose the gallows humour, as this is how we recover from unpleasant incidents. I said I basically disagreed with that, and said that this is all connected,
the culture of banter, joking, minimizing things through comedy is all part of
an unhealthy culture and it should all change. Bantering, joking, doesn't really
deal with problems properly. The dark, joking culture is part of a 'man up' culture which is wrong. You hear of this still: 'the best thing is just not to think about
it at all, just block it out'. That's unhealthy.

Discussions of diversity and equality fit the broader sense in which ambulance services have gone through major but incomplete cultural and
operational transformations. Radical changes and improvements coexist
alongside areas where there has been little or no change in customs, practice, and conduct. Overall, there is a strong sense that the 'old school'
culture that dominated before the rise of the professionalization project
is in significant decline. The ambulance service is no longer the 'monoculture' that it once was. It has proceeded a long way from the 'blue-collar
professionalism' of Metz (1981). Nevertheless, some very strong cultural
resonances have survived into the current era, most notably the persistence
of roadstaff-manager conflict and mistrust, a climate of work overload
and strain, an informal value hierarchy of callouts which privileges 'genuine emergencies' over unplanned primary care, and a persistent sense
in which ambulance crews will almost willingly seal themselves off from
other groups. Their cultural practices sometimes generate a kind of siege
mentality that may be mutually supportive in limited ways, but contains
weaknesses and drawbacks of its own. These cultural elements are durable
and understandable reactions to the intense difficulties associated with the
paramedic role, a role that is growing in professional status yet continues
to confront severe challenges and limitations.

Conclusion

This chapter has explored what I understood to be the 'daily working
culture' of England's paramedics. It has included various viewpoints and
dimensions of the paramedic role, especially how they feel about the ways
in which paramedics are 'put to work' by their employing ambulance Trust.
Professionalism has an important presence here, but in a somewhat attenuated, muted fashion. The intense pressures of patient demand weigh

heavily on everyone working in England's ambulance trusts. There is relentless pressure to try to meet this demand and the arrays of performance measures that dominate the managerial agenda. The managerial regime was understood by paramedics in peculiar ways. For some, they would describe powerful ways in which 'management' would exercise control over them. Other paramedics would emphasize their considerable autonomy, and would suggest that management leaves them alone, or even ignores large areas of their actions and practice. Some described NHS Trusts and their management as indifferent to paramedics' emerging sense of professionalism, even as the idea of being 'professional' was of major individual and collective importance.

There were tensions over the increased degrees of risk entailed by the enhanced discretion and scope of practice associated with professionalization. Some embraced the discretion and autonomy more than others, with the younger paramedics (most of them Paramedic Science graduates) usually keener to characterize the paramedic role as a clinical one, embracing free thinking and autonomous practice. Complaints about management were common across all roadstaff, seemingly regardless of their background, reflecting not only long-held frustrations over call prioritizing and uncivil relations with managers, but also a growing new sense in which some paramedics feel they are outgrowing their employers. Many of the managers were described as exhibiting 'old school' traits that the frontline, especially degree-trained, paramedic professionalism are leaving behind. On the other hand, the incessant criticism of management and control could also be exaggerated, and there was an important sense in which 'the moaning' was a cultural artefact that 'came with the territory' and actually re-enforced the supposedly outdated 'old school' cultural features that 'clinical professionals' were supposed to shun.

There were also some interesting tensions around the roles and responsibilities of patients calling 999 or 111. The nature of the calls and the nature of the broader service are changing. Some paramedics were fairly comfortable with the changing composition of their callouts. While there would always be complaints about inappropriate call prioritizing, there was a growing acceptance that it is wrong to consider only major trauma and cardiac emergencies as the 'real' work of paramedics, and that calls to patients suffering from anxiety, falls, or chronic illness are somehow a waste of time. Low acuity callouts still usually meant patients in need of care of some kind. Urgent primary care calls can also provide room for more clinical discretion. Trauma calls often require advanced skills and risky procedures, but, paradoxically, they are also often fairly straightforward in terms of decision-making. Life-threatening emergencies usually have clear pathways and procedures to follow under considerable time pressure. These features are all much less prominent with callouts to patients suffering from chronic problems such as severe depression, alcoholism, and dementia. While low on the hierarchy of traditionally valued calls, these complex 'social care type' calls, often not involving a run to hospital, are ones in which paramedics can enact compassionate care (Seim, 2020: 78), and exercise discretion, workarounds, improvisation, and negotiation with other stakeholders, in the traditions of street-level bureaucracy.

Professionalism, therefore, is not only found in the ability to deal effectively with clinically demanding cases involving critically ill and injured patients. Another part is being able to handle the complex negotiations involved with the grey areas surrounding low-acuity 'social welfare' type calls, to be an advocate for every patient encountered, to figure out a better path forward so that the patient does not have to resort to emergency callouts again. Some of these calls would be mentally written off as 'rubbish' and 'crap'. But another part of a 'professional' bearing is to bury these feelings, maybe to allow them to come out in private or in backstage areas, but never around patients and family. A 'professional' ambulance responder might engage in 'the moaning' but can't let it take over. They will block this out. 'It is what it is'. Constant moaning is a waste of energy. A kind of stoicism takes over, a kind of mind over matter, an element I will return to later in the book. This is not to say that this is somehow easy. Ambulance work is extraordinarily challenging. Stress, PTSD, depression, burnout, and illness are chronic problems facing everyone working in the service. These stresses, strains and coping mechanisms are the topic of the next chapter.

5

Coping with Unbearable Strain

> '[W]hen the desire for wholeness leads to an enlistment in greedy organizations, it may end in an obliteration of the characteristics that mark the private person as an autonomous actor.'
>
> **(Coser, 1974: 18)**

Paramedics and other ambulance responders face acute and chronic work-place stressors. There is an extensive academic literature on paramedic stress, psychological injury, and burnout, drawing on concepts such as psycho-social strain, psychological safety, resilience, stress, PTSD (post-traumatic stress disorder), and moral injury (Clompus and Albarran, 2016; Maguire et al., 2014; Mildenhall, 2012, 2021; Murray, 2019). Paramedics' exposure to emotional and physical risks is a well-known international problem (Lawn et al., 2020; Nelson et al., 2020). Several national governments have commissioned inquiries and reports into this matter, with one of the most extensive being the Australian Senate's *The People Behind 000* (Senate Education and Employment Committees, 2019).[1] Much has been written about paramedics' 'coping mechanisms' (Clompus and Albarran, 2016; Henckes and Nurok, 2015), which are often portrayed as similar to those working in other 'extreme' settings (Granter et al., 2015; Hällgran et al., 2017) such as emergency room doctors, police officers, firefighters, or combat veterans (Frankfurt and Frazier, 2016; Gabbert, 2020; Pogrebin and Poole, 1988; de Rond and Lok, 2016). Some stress reactions can be ameliorative: reflection, informal discussions with colleagues, taking time away from work, therapeutic counselling; whereas others are usually harmful: withdrawal, problem drinking, and prolonged feelings of anger, shame, worthlessness, and guilt.

Laypersons often assume that paramedics attend shocking and traumatic scenes on a daily basis. This perception may derive from cultural

---

[1] Related reports in Britain have included the 'Blue Light Report' from the mental health charity Mind (2018), a report entitled 'In Harm's Way' by the GMB trade union (GMB, 2018), and governmental inquiries on ambulance services by the National Audit Office (2017) and the UK Parliament's Public Accounts Committee (House of Commons Public Accounts Committee, 2017).

The Paramedic at Work. Leo McCann, Oxford University Press.
© Leo McCann (2022). DOI: 10.1093/oso/9780198816362.003.0005

signals broadcast in TV and film, hence the media characterization of emergency medics as 'heroes', and the often-heard refrain from bystanders that 'I couldn't do what you do'. The reality is more complex. To some extent, it is true that exposure to stress 'comes with the territory' of ambulance work. Few forms of employment require staff to deal with such exhausting and high-stakes activities as driving on blue lights, treating life-threatening or life-changing injuries, attempting to resuscitate critically ill and dying patients and dealing with the emotional demands of distressed family members and bystanders. All of these tasks can be highly stressful, with ambulance staff at risk of suffering from post-traumatic stress disorder and other psychiatric injuries. The prevalence of PTSD symptoms among the worldwide paramedic community is estimated to be around 11%, anxiety and depression around 15% and general psychological distress around 27% (Petrie et al., 2018). These figures are well above those seen in the general population (Mildenhall, 2019). A recent survey by the trade union GMB (2018) suggested as many as 39% of UK ambulance staff have experienced PTSD. There are good reasons to interpret all this as amounting to a mental health crisis in England's ambulance services.[2] Stress, therefore, does indeed 'come with the territory'.

But part of the discourse of ambulance work being unavoidably stressful stems from the common misunderstanding that emergency ambulance work is a non-stop series of road accidents, falls from height, suicides, stabbings, and heart attacks. Obviously paramedics do respond to these kinds of emergency but, for most responders, such calls are not experienced frequently. A small number of paramedics in specialist roles, such as CCPs, RATs, or those working on an air ambulance, do face life-threatening sickness and trauma on a daily basis. But for most paramedics, calls to very serious trauma or life-threatening emergencies such as cardiac arrests and strokes might occur once every 6–7 shifts (as a very rough estimate). And we would be wrong to assume that these cases necessarily result in emotional trauma for the responder (although they may do). Truly terrible events that are dramatic, unforgettable, and potentially overwhelming (such as multi-casualty accidents, dangerous and violent scenes, or incidents necessitating investigations and giving evidence in court) are encountered much less commonly, probably around every 3–4 years. At times I listened to paramedics mention two or three critical incidents they have encountered at earlier points in their career, from being

---

[2] 'Revealed: The hidden crisis in Britain's ambulance services', *The Spectator*, 30 August 2014.

unable to shake intrusive memories of a terrified young man dying of stab wounds, to the feelings of being overwhelmed at major incidents of national importance such as train crashes or terrible acts of violence.

There are complex intricacies involved around how responders might react (or not) to the specifics of the various scenes they encounter. People can become upset, challenged, or haunted (or not) by different events, and sometimes not by those one might anticipate (Maunder et al., 2012).[3] Sometimes paramedics become chronically stressed by the slow build-up of the general sadness of working amid sickness, suffering, and loss or by the relentless pace of callout demand. Others are upended by a particularly gruesome, frightening, dangerous, or distressing call. Sometimes a stress reaction is hard to explain. Several paramedics recalled incidents where they became upset by the experience of working with patients who reminded them of loved ones, often their parents or children. It was very common for paramedics to mention that 'anything to do with kids' are the most difficult kinds of call. Less common, but equally forceful, was the traumatic possibility of attending a scene involving the injury or death of a family member, friend or colleague. Responding to an emergency involving injury and death to children, relatives, friends, or fellow emergency workers were probably the worst scenarios imaginable, and these terrible subjects weren't talked about to me in much detail. Brief mentions of such incidents would take place in a less candid way than almost all other subjects, serving as an unspoken 'marker' that further discussion of this area is off-limits to an outsider like me. Certain areas of ambulance work are, therefore, quite rightly understood as forms of 'extreme work' (Granter et al., 2015).

What is perhaps less well known is that the causes of stress and strain for paramedics are broad, complex, and interlocking, going well beyond the most obvious stressors of exposure to traumatic incidents. Previous chapters have already shown that paramedics are, to some extent, accepting of the risks associated with responding to life-threatening calls, that many of them actively seek them out, and that an informal status hierarchy pertains to these serious calls and to persons seen to deal with them appropriately. It is instructive that these potentially traumatic calls are invested with such meaning. Although they have no patience with the 'hero' narrative, paramedics do have a stoic acceptance that death, pain, loss, and fear are elements of the work that they have 'signed up for', that they have to be

---

[3] 'I've seen terrible things as a paramedic. The worst isn't what you'd expect', *The Guardian*, 5 April 2018.

ready to confront all this effectively and to be able to successfully 'process' the after-effects (Furness et al., 2020). This can be problematic. The need for resilience and toughness in the face of extremely trying events tends to reinforce stereotypical 'macho' traits of uniformed occupations, meaning that responders might struggle to 'admit' that they are suffering mental distress and will decide not to seek support (Mildenhall, 2021). The term 'burnout' did not really exist in the early 1970s, but Coser's description of the possibilities for the 'obliteration' of 'an autonomous actor' speaks powerfully to the intense risks that accompany a person's 'total commitment' (Coser, 1974: 18) to the identity, role, and demands of the paramedic. The professional field of the paramedic enjoys a high level of social respect and features some powerful and unusual rewards. But its tendencies towards extreme work and edgework, its all-encompassing operator culture, its deep social meaning, and the endless operational demands mark the field of ambulance work as a 'greedy institution' (Coser, 1974) that makes huge demands on its occupants that can be temporarily or permanently overwhelming.

To some extent, the penalties of emotional burnout and moral injury are partially offset by narratives of ambulance work being 'the best job in the world', 'what we do', 'a privilege', or 'what we're trained for'. One ambulance technician explicitly asked me not to 'only report our miserablist stories'.[4] Working in this field entails a complex tangle of emotions. Handling sad, traumatic, and unusual incidents is a crucial element of the social meaning of the profession. But this need balancing against the high risk of physical and psychological injury that paramedics face (Maguire, 2014). The risks of burnout and moral injury are the flipside of the 'privilege' of being a 'witness to suffering' and an advocate for patients in need.

On top of the risks associated with their irregular attendance at traumatic emergency scenes, paramedics also face other, chronic stressors on a daily basis. These might be considered 'background' stressors, which come in a constant 'drip feed' of smaller and more mundane issues, often building up to a point where ambulance work becomes harmful to health. These background stressors derive from the relentless operational pressure, public misuse of the service, questionable call prioritization decisions, unsupportive line management, the fear of making mistakes or 'missing something', and the lurking presence of complaints and disciplinary action. The

---

[4] 'The NHS might be at breaking point, but is there still room for optimism', *Huffington Post*, 26 July 2018.

working life of a paramedic, while in many ways unique, also shares a range of mundane stressors familiar to anyone working in a large organization: faceless bureaucracy, poorly designed rules and procedures, exhausting shift work, resource overstretch, unsupportive management, time wasted on meaningless activities and strained interpersonal relations. One metaphor I came across (which seems to have originated in Australia), was that of 'the bucket'; everyday 'drips' of exhaustion, conflict, and suffering gradually fill the bucket to its brim, at which point all it takes is any matter, large or small, for everything to spill over.[5]

The professionalization agenda has had some success in changing and improving many aspects of paramedic working life in England, especially through the expansion of education and clinical scope. Professionalization has also enabled paramedics themselves to research and campaign about aspects of their work including clinical autonomy, and health and wellbeing, contributing to the opening of a space where mental health is being much more widely discussed. But there are distinct limits as to how this plays out in practice. Operational demands mean that there is little time for formal and informal support practices, with managers also overloaded and stressed. Staff 'buckets' were said to be at or near capacity pretty much all of the time. An absence of supportive management is probably the major reason why paramedics leave the ambulance service.[6] News articles published even before the Covid crisis described NHS staff 'in a very deep hole',[7] with patients and family members describing A&E units as 'utter mayhem'.[8] Long waits for ambulances were described as 'beyond a joke'.[9]

This chapter delves into the most unpleasant and troubling elements of paramedic work, from the stark realities of operational overload, the uneasy realities of 'coping mechanisms' and to the prevalence of ill health, mental trauma, and even suicide among the paramedic community. It explores the enormous pressures placed on those working in the service, in which workers risk having their wellbeing 'slowly scratched away'. It documents how ambulance workers and employers try to handle the

---

[5] 'Inquiry into mental health of emergency service workers hears of "the bucket"', ABC News, 31 July 2018.

[6] 'Factors affecting the retention of paramedics within the ambulance service', College of Paramedics, 21 May 2019, available at: https://www.collegeofparamedics.co.uk/COP/Blog_Content/Factors_affecting_the_retention_of_paramedics_within_the_Ambulance_Services.aspx (accessed 13 July 2021).

[7] 'NHS winter pressures: Doctor says staff in 'very deep hole', BBC News, 31 January 2020.

[8] '"Misery" for A&E patients facing record-long waits', BBC News, 9 January 2020.

[9] 'Seriously ill wait more than an hour for an ambulance', BBC News, 29 January 2020; 'Ambulance service: Waiting system "beyond a joke"', BBC News, 29 January 2020.

impact of distressing calls through formal and informal means. Alongside the gallows humour that famously operates in these, and other, 'extreme' settings (de Rond, 2017; Gabbert, 2020; Granter et al., 2015; Hällgran et al., 2017; Pogrebin and Poole, 1988) serious emergencies also provide opportunities for learning and debriefing, and they occupy a high position in the informal status hierarchy of paramedic work discussed in Chapter 4. However unpleasant they are, I show that very serious and challenging calls are part of the essence of emergency work, and that competent and 'professional' paramedics have to find appropriate ways to deal with them. The chapter moves on to explore the limits of coping mechanisms, drawing on academic concepts such as burnout, PTSD, and moral injury, but also the folk concepts emergent from paramedic street culture used to describe emotional labour, such as 'five-year job', 'the bucket', and the 'grief mop'. Finally, it also describes endemic problems associated with ambulance service managerial and employment cultures. These represent yet another form of stress for paramedics, who often feel undervalued and neglected and sometimes live in fear of accusations of wrongdoing, complaints, and inspections.

What follows can be tough reading in places. My understanding of the paramedics' orientation to this subject was that they wanted their mental and emotional challenges to be recognized, especially when it comes to the urgent need they perceived for ambulance trusts to change their operations and culture in ways that would better address the crisis in paramedic wellbeing. I felt they wanted me to portray many of the grim realities. But they didn't want to be portrayed as tragic figures, pitiful, put-upon, and victimized. Throughout the study, they repeatedly returned to much more optimistic and rewarding themes of the 'privilege' associated with serving the public, the social and moral importance of helping others in need and the rewards associated with the expansion of paramedic professionalism so that patients can be better served. I will return to these more optimistic themes in the following chapter.

We start this chapter with a discussion of the challenges associated with the intensity and pace of ambulance work in conditions where callout volume continues to climb ever upwards.

'We Deal in Chaos': The Nature of Demand Overload

Prior chapters have repeatedly mentioned the steady rise in demand for ambulance responses. In this section, I explore how this relentless demand can impact on paramedics' health and wellbeing. Given the chance, it is a topic that nearly every ambulance worker will readily discuss in some detail. On a shift with CTL Chris,[10] for example, I learned a great deal about the patterns of calls in this specific location; the peaks and troughs in demand, the long-term increase in callout activity, and the Trust's managerial struggles in trying to design and adapt its systems to accommodate it all. Chris outlined the various geographic and cultural differences within the Trust, where call activity is most concentrated, which stations have the busiest crews and which stations have a reputation for being quieter than others. In the end, his view was simultaneously philosophical and resigned: 'There's no rhyme or reason. We deal in chaos.'

'Dealing in chaos' was a crew-level perspective. Managers at more senior levels would, of course, take a different view, pointing to complex analyses of demand trends, deployment plans and preparedness policies. All NHS ambulance trusts, for example, operate under the 'REAP' levels, of: '1 – steady state, 2 moderate pressure, 3 – severe pressure, 4 – extreme pressure', and each organization involved in ambulance response generates reams of publicly available policy documents about 'resilience' and 'preparedness'. Crews at the front line are familiar with these concepts and acronyms. But they would tend to talk in much more colloquial terms about patient demand, using such phrases as 'rushed off our feet', 'moved from pillar to post', and 'let's see what the job is when we get there'.

The notion of 'chaos' also appeared in a discussion with a group of paramedics and managers who had attended a presentation I gave at a CPD event. Kevin, a paramedic team leader very much with an 'old school' bearing, gave his views as follows:

> It's chaos a lot of the time. People from outside don't know this. There are so many gaps in the coverage.

'Gaps in the coverage' related not only to there being too few ambulance vehicles available, too little capacity at A&E units leading to ambulance queues and handover delays, and too few qualified paramedics. Ambulance staff would also point to a lack of adequate clinical cover from other healthcare providers and occupations. Observing on a shift with Pete, he mentioned a growing problem with the volume of suicide attempts that he and his partner have responded to in recent years. His account begins with familiar criticisms of a lack of resource and lack of management attention, but he goes on to situate his critique within a broader story of a general

---

[10] All names of people, places, and organizations are pseudonyms. For details on the research process involved with the writing of this book, please see the Appendix.

absence of care availability, referring not only to the crisis of in-community psychiatric and social care, but also to what he saw as a crisis in society more generally. Communities are scattered and isolated. People lack access to basic help from families and friends. Social problems now develop into 111 or 999 calls when, in prior times, 'the community' itself could have handled them without asking for assistance:

> On the subject of suicide attempts, Pete says the psychiatric cover is 'non-existent'. 'Out of Hours, in particular mental health care, is horrendous. And it's very hard for us. You take them to A&E and sit 'em down and they get up and leave the waiting room. Even when we call the psych crisis teams, there is no follow-up. The thing is, family used to handle these sorts of things. Everyone's now so spread out, there used to be more support locally. This has all dwindled for whatever reasons. You don't know who is living near you, there's no community left.'

Pete's notion of the general public being spread out thinly seemed to mirror the sense of overstretch that paramedics often feel in their working lives, in a dual sense of both the long geographic distances between many of the calls and the chronic lack of adequate coverage. Long hours behind the wheel, and the constant 'grind' of work is causing people and equipment to break down, in many cases literally. Chris claimed that many of the ambulances in use in his Trust had 200,000 miles on the clock after four and a half years, and that it was not uncommon to hear of vehicles with 500,000 or 600,000 miles on the clock after six years. Staff would talk of 'thrashing the life out' of the road vehicles they used.

Covering a lot of ground in the pursuit of response targets led to feelings of dislocation, often described using strange and surreal imagery. In a detailed reflection of night shifts, tiredness and the sights and sounds of ambulance work, David describes in an interview the 'ping pong' and 'treadmill' associated with the 'river' of constant jobs. He hears sirens, alerts, tones, and bleeps from the MDT system in his sleep. His words meander through a haze of tiredness, at times having an almost hallucinogenic quality, with the ambulance at one point portrayed almost as a meat wagon or 'butcher's shop':

David: Working in Metroville was really a treadmill, like ping pong. They'd be a job almost always, you'd climb back in the vehicle after you clear off the prior job at the hospital and the MDT would immediately give you another job – the little alert it makes is like a siren: 'wah-wah!' You'd hear this in your sleep. It could be overwhelming. It was like a surge, a surge of jobs, just waiting beyond you, press the button and in comes the next one, like a river. [. . .] I sometimes found the sights and sounds weird and they would stay with you. With the Lifepak, that would usually stay in place on the wall of the ambulance, and we'd often loop the leads up and over the handrails near the ceiling of the ambulance. The BP cuffs would hang down and they look like rats' tails, or something weird hanging down, like in a butcher's shop.

Shift patterns, long miles on the road and the frequent movement of location are tiring and disruptive. As an observer, I made the following fieldnotes the day after finishing a 12-hour shift, a completely ordinary one in which all of the calls were routine and low acuity. I noted that I still felt drained the day after the shift, and that's with me just observing and taking notes, not doing any clinical work, driving or manual handling. I recall a specific form of tiredness I'd not felt before, a feeling of fuzziness, not unlike jet lag:

> It was a very routine, steady shift, but still I feel very tired afterwards. I noticed that at about 4pm the day after the shift I felt wobbly and lightheaded through lack of energy. I'm very tired and drained. Often my eyes seem to want to close by themselves. I feel tired across my whole body, I'm not wanting to do much.

This experience provided just a glimpse of how 12-hour ambulance shifts might make a person feel. I struggle to imagine that experience repeated for a string of shifts across several days.

The idea of 'chaos' was one way that the responders evaluated what they do. It was perhaps a way of shrugging away some of the pressures. The overall structure is 'chaos'. There are delays. There are stupid calls. The equipment is 'thrashed'. Everyone is 'knackered'. You can only do what you can. You are doing your best and your best is all you can give.

'Slowly Scratched Away'

The repeated, continual grind of tiredness and frustration is a major risk factor for the health and wellbeing of paramedics. This is clearly reflected in ambulance trusts' very poor records for staff sickness absence, being by some distance the worst of all types of NHS employer (McCann et al., 2015; Nelson et al., 2020). Paramedics in 2019 took over 52,000 days off work due to psychiatric illnesses such as stress, anxiety, and depression.[11] Always a problem, this issue has, not surprisingly, worsened since the Covid epidemic broke out in 2020. Simon described this to me using the kind of stark, metaphorical language that he favoured:

> Simon: 'You don't get broken by one bad job. It is a sum total of erosion, you get slowly scratched away.'

The general, long-term strain or 'grind' of ambulance work is interesting in that it does not refer to any specific incidents that might directly trigger emotional distress. Rather, it refers to chronic, background stressors, leaving paramedics with few obvious 'coping mechanisms' to mitigate them. It is this background pressure that stimulates the endless chatter on station about rosters and holiday bookings, and the frequent moaning and griping, just like in many other occupational settings.

Beyond this, paramedics—like other 'street-level' (Maynard-Moody and Musheno, 2003) workers serving the public in complex contexts—have developed sophisticated techniques for regulating their emotions and handling the fallout of emotion-laden work (Boyle and Healy, 2003). This was often described as developing 'a protective shell', often through various forms of 'distancing'. It was common to learn of the dangers of 'getting too close' to patients. This bearing was similar to the self-control about 'the moaning' discussed in Chapter 4. Paramedics needed to control their emotions, find ways not to get overwhelmed by them, a bit like the way they must also not give in too frequently to the urge to keep feeding 'the moaning' machine. At times, some roadstaff would project an almost monkish, soldier-like asceticism. Recall the paramedic in Chapter 4 using the metaphor of 'adapt and survive'. One aspect of this involves 'bottling up' some of the emotional trauma they have experienced. Another is the widespread use of 'distancing' type behaviours. David made some interesting comments on this, drawing on the ways in which paramedics were educated and trained. 'Everything is a simulation', he suggested, 'like a game or a puzzle':

> David: Everything is slightly removed – training on manikins, you know, plastic people, its simulatory. Everything is a simulation. It's like a game or a puzzle. This then feeds into your own coping mechanism. Sometimes you also see the

---

[11] 'Mental health leave for paramedics in England nearly triples since 2011', *The Guardian*, 23 July 2020.

real patients like a simulation, you withdraw from showing real empathy. The empathy often isn't real, it's like 'there, there', while you are focusing on the clinical issue. Is your thinking clear? Don't let yourself get too drawn in on any emotional levels or overly emotional behaviours. The uniform is kind of a shield. Sometimes you think these symptoms aren't real, they are from the person's mind, and you have to sort of block them at times. [...] I struggled with this at first as I got too close to people's concerns. You can't feel everything for everyone. But then again, they have a life story, and you often hear significant amounts of it. So, someone will tell you they had a hernia op 13 years ago. You are a witness to their suffering. You are that witness, and you have to learn not to resent them. You do get exhausted.

Several ideas float around in this passage of speech. It seemed as if David's experiences were a mixture of things he had learnt by himself, when he 'struggled with this at first', combined with other practices, ideas and metaphors that circulate the profession as a whole. The uniform as 'a kind of shield' is very reminiscent of observations that appear in the ambulance memoirs. Here, for example, is the highly experienced, now retired, paramedic Lysa Walder (2020: 177–8), using similar military and defensive metaphors alongside more typically NHS-style ideas around emotion management:

The seen-it-all image, the protective shell most paramedics don, along with our uniforms, our stab vests – our armour if you like – must be maintained at all costs. [...] You couldn't do this job for long if you gave into your own feelings all the time.

These comments accurately describe some of the emotion management techniques that ambulance responders develop. They reveal the complex emotional interplay between the various ways that paramedics interact with patents. Sometimes they use, in Seim's words 'cold hands' and at others 'warm hands' (2020: 80). Too much coldness and the paramedic risks appearing uncaring and 'cynical'. Too much warmth and they've 'got too close' to their patients risking compassion fatigue.

David's viewpoint also hints at an important aspect of street-level professionalism in that paramedics have chosen this occupation and therefore have limited licence to complain about how tough it can be. As professionals their job is to put patients first, which means being competent to handle the nastiest things they encounter on the streets. Paramedics have a licence to moan about poor management, failing equipment and rubbish calls, but trauma, suffering, and death are closely associated with the core meanings of ambulance work and need to be accepted as such. Just as paramedics can sometimes exaggerate with their complaining about management and patients, there is a danger of exaggerating how hard the job is. David spoke powerfully about how paramedics should never portray themselves as victims. Returning to that core idea about ambulance work as 'a privilege', the pain and suffering felt by the public in crisis should not be usurped by those of the emergency responder. The patient and the grieving family members are the ones being victimized by accidents and illness, not the responder. David and others would regularly point out that most paramedic work is routine, and that much of it is enjoyable. Much like the complex issue surrounding what are 'genuine' emergencies or 'worthwhile' calls, some of the incidents of mental ill health and burnout faced by paramedics are real, but others might be derived more from the result of the ambulance service's troubled employment culture, rather than from 'genuine' trauma and stress. David mentioned a terrible incident that would have 'qualified' for him to 'go out with stress', had he wanted to. But he chose not to, and his reflections on this on the broader meaning of ambulance work are instructive.

> David: I wonder sometimes about the reaction when people say they are sick with stress and go out with stress. Sometimes I think this is also basically a reaction against management. I mean, I've certainly been to jobs here and there that have stayed with me, where it's been a very upsetting scene that keeps returning to your mind. Sometimes it's about when there's a so-called Act of God, and it is just horrible, just terrible bad luck and it is unjust and has no meaning. There was one in particular [. . .] A young woman was just walking along the street and there was construction nearby, and a whole pile of construction materials just crashed down onto her. She had multiple severe injuries and there was no chance of saving her. One moment you are going about your business, walking down the street, the next moment you're horrifically injured or dead. For no reason. For some months I kept seeing elements of this job in my mind's eye. Sometimes you wonder who she was, how her family has been hit by it. But there is nothing you can do. So eventually it does go away. I mean, I'll never forget it, but it no longer troubles me. Now I could have gone and asked for counselling, or said I'm sick, or stressed or PTSD or whatever. Got some time out maybe. But what's the point? I'm ok. I will get through it. She's the one affected by the tragedy. Her and her family. It wouldn't be right to say this has

made me ill – it hasn't, I'd just be using this to get time off, again, sort of venting at management.

For all the talk of stress, strain and extreme work, one particular interviewee even described work as a paramedic as 'not that hard', suggesting 'most of the jobs are easy'. This was the view of Liam, a very experienced, academically highly qualified paramedic. I feel he is exaggerating a little. But, in light of the nature of calls and the operator culture described in Chapters 3 and 4, the overall point he makes is largely valid:

> Liam: They'd be some staff who moan even if they were given a million pounds and endless extra time for training. Some staff, sadly, are now so disillusioned they shouldn't really be working in this field. But where do they go? They don't have many qualifications, they have the IHCD, technician qualification, or NHSTD, it's like the grandson of the Millar's Certificate. They're stuck with a certain mentality, a certain way of working, and no real qualifications. They can't move on. If you hate it so much you should leave, but where do you go? If you're fed up with the twelve-hour shifts, the grind, you should really get out. Ok, it is exhausting at times. But is it really that bad? It's not that hard. In fact, most jobs are easy. And the banter, the interaction with the public is mostly good. You hear horror stories, but mostly it is good, and the good people, you are respected, they do say 'thank you'. Once you're at a job, you're there, you can't get another job. It's a bit different on the cars, but on an ambulance you'll get one job per hour, you handle each job as it comes up, and there it is. Some jobs are hard, stressful, for sure. If you get a really bad trauma, a paediatric job, or a childbirth that is not progressing, but these are relatively rare. In a way sometimes that makes them harder to deal with emotionally. Or you worry about your skills fading. That you've not kept up on maternity, or whatever, and then you get a bad maternity job.

As his account goes on, he starts to complicate it by including the risks and difficulties that circulate 'jobs' that are challenging and risky but unusual. Paramedics want to use their skills, and they want to help the public. That means they value high-acuity calls. They want their work to be meaningful, and they are bored and frustrated by 'rubbish' calls (Corman, 2017; Mannon, 1992; Metz, 1981; Palmer, 1983; Seim, 2020). On the other hand, they would struggle if confronted by a non-stop string of 'really bad trauma calls'. They are also very mindful of the risk around 'a skills drain' when they feel unprepared for a serious or unusual call that can appear at any time.

There are limits to what anyone can deal with, especially when exposure to stressful incidents is inevitable, expected, and repeated (Grayling, 2017: 171–84; Shay, 1994).

Some paramedics were very candid about how the work had made them ill, sharing details about times when they have been off work or diagnosed with stress-related illnesses. Some had accessed counselling, which they usually described as welcome and effective. While a difficult area to research, the observation phase of my research meant that I could not avoid it. Discussions on station often featured talk of who was available for work and who was off sick. People sympathetically asked around for the latest news about staff who weren't currently available. I came across various upsetting examples, such as one responder who had been off work and undergoing TRiM counselling due to flashbacks from attending a terrible and deeply confusing scene involving life-threatening violence.

My interview with Karen took place while she was off sick, apparently because of chronic stress:

Karen: I had an episode of SVT.[12] I just got dizzy and ill at a patient's house. It turned out to be SVT – Ok, that's it, time out! I've had three weeks, but I'm back next week. [. . .] It's just a slow build-up, a combination of things over time. There's a lot of stress around. Some of it includes PTSD symptoms. Sometimes acute, but often chronic. It's just continuous and there is no break from it.

I asked her if ambulance staff are starting to need more official interventions, perhaps forms of therapy and counselling:

Karen: Yes definitely. It's definitely reached that point. I decided I needed some help. I need to learn how to manage my stress better at work. The majority of issues I face are beyond my control. But this is a hard thing for me to accept, really. I want to fix things, I want things to work. But I feel powerless and frustrated. Many of us do.

We are back—yet again—to impossible demands and resource overstretch. The pressures the NHS faces as a whole are being projected onto the shoulders of its employees. The resulting effects on staff health are obvious. More subtly, there are other damaging effects, too. Paramedics often felt that the service they provided was inadequate. The long waits, the lack of

---

[12] Supraventricular tachycardia—abnormally fast beating of the heart.

time set aside for training and development, and management's obsession with performance metrics all contribute to a slowing, even a reversing, of the progress of paramedic professionalization (McCann et al., 2013). For many, this tough reality preyed on their already low sense of self-worth. When patient care is regularly compromised by long waits, perhaps especially by long secondary waits, then a responding paramedic might feel strong emotions of powerlessness and guilt. For all the 'armour plating', the distancing, and the decompressing, paramedics run the risk of feeling complicit in the ambulance service's 'failings'. Here we have the prospect of 'moral injury' (Murray, 2019), a concept that, alongside PTSD, emerges from the protracted trauma, violence, and chaos of war (Shay, 2014). Jonathan Shay (1994) writes of the 'undoing of character' in warzones, where the moral value of traumatic and surreal acts and events are impossible to rationalize or understand (see also de Rond, 2017: 85). There are strong overlaps between many of the experiences of emergency medicine and of warfare, a point not lost on practitioners (de Rond, 2017; Seim, 2020: 114) and commentators, for example where overcrowded English hospitals were described by the British Red Cross as facing a 'humanitarian crisis'.[13]

Although NHS pre-hospital work is not directly comparable to the horror of battlefield conditions, there are rare occasions where what the paramedics have seen (and done) offends their professional judgement and moral calculus. Moral injury can result when an emergency professional feels that the service they have given to patients and family members has been inadequate, and where patients and their families had been let down, often by confused interactions at a complex incident, or by unacceptably long waits for an available ambulance to arrive for onward transport of a critically ill patient. For all their often-expressed cynicism, I felt that some of the paramedics seemed to take on too much responsibility for issues that were beyond their control, and that they felt the injustices of the system perhaps too keenly. This is understandable when placed into the context of the unique way that ambulance response works. There is a major downside to something that is often seen as an advantage of ambulance work—the fact that the ambulances usually respond only to one patient at a time. While paramedics greatly value and appreciate being able to have one patient as the sole focus of their clinical attention, this situation also poses a distinct disadvantage. Where there are only one or two responders

[13] 'NHS faces "humanitarian crisis" as demand rises, British Red Cross Warns', *The Guardian*, 7 January 2017.

present, they become isolated as the 'face' of the ambulance service and, to an extent, the NHS as a whole. They confront a danger, therefore, in being saddled with responsibility for problems beyond their control. In some sense they embrace this. Like Karen, they 'want to fix things'. One interesting research study suggests that paramedics are more likely than other categories of the general public to have experienced early life abuse and neglect, implying that they might have especially highly attuned senses of empathy and injustice (Maunder et al., 2012). I found that their reflections included an intense, ethically infused focus on patients in real need: 'witness to suffering', 'a privilege'; combined with a cynicism about those less worthy of an ambulance response: 'rubbish', 'crap'. Paramedics operate in a morally complex and emotionally laden environment (Boyle and Healy, 2003) where the options they have are often compromised, where they face difficult judgements and where they cannot be certain that they have done the right thing—a situation quite similar to many other public sector and uniformed settings (Hupe and Hill, 2019; Kaufman, 2019; Zacka, 2017). The risk of moral injury is ever-present.

Many staff find the combined, interlocking pressures and risks unbearable, and have quit ambulance work altogether. As a researcher interviewing and observing currently employed paramedics, this was something hard for me to get a clear angle on. Several told me they were thinking very seriously of leaving the service to do something else. Something I began to hear in the last year or so of my interviews was the notion that paramedicine had become 'a five-year job'. The idea that a paramedic career will now 'last' only five years is tragically unreasonable given the requirements of the three-year degree for entry. The 'five-year job' is a powerful notion, but seems to have little basis in fact. It reflects a sense in which the 'new breed' of degree-educated paramedic now has the qualifications and perhaps the mindset to act on desires to leave, an option less viable for those trained in-service. Paramedics are outgrowing their employers. Some do leave NHS trusts in search of something better. But many do not, finding ways to balance out the stresses and rewards over many years. Without wishing in any way to downplay the extremely challenging nature of ambulance careers, the 'five-year job' is a good example of an exaggerated notion that often becomes part of roadstaff folklore, something that is not in itself quite true, but is an instructive reflection of genuine grievances and struggles.

Although paramedics take pride in their role and often describe their job as worthwhile and rewarding, their accounts of the broader self-identity of the paramedic profession can often be strangely self-deprecating. They refuse to accept the popular 'hero' narrative, and they endure a difficult workplace culture. This feeds into the development of defensive forms of resilience that take on the flavour of 'them and us', and 'armour plating.' These siege-mentality style coping mechanisms have some value, but also contribute to the risk of paramedics being unlikely to seek help when facing a wellbeing crisis, and makes them potentially vulnerable to feelings of low self-esteem, powerlessness, and worthlessness. Such a scenario could hint at some of the reasons for the disturbingly high prevalence of suicide among paramedics, a truly dire situation that several ambulance staff mentioned to me, and a problem that seems to be becoming increasingly serious. Mars et al. (2020: 11) find that the suicide risk for male paramedics is 75% higher than the national average for males in the UK. At one point in late 2019, one particular English ambulance trust experienced three staff suicides in 11 days, with allegations circulating of 'toxic bullying.'[14] Government statistics suggest that 42 paramedics died by suicide between 2011 and 2019.[15] It is not unusual to read desperately sad Twitter posts mentioning and commemorating paramedics who have died in such circumstances. These posts often using language not unlike that of the police, in which 'another member of the green family' has been lost. The local effects of a staff suicide on a particular ambulance station can be devastating. Deep wells of tragedy and sadness surround the harsher edges of the ambulance world.

'If You Don't Laugh, You Will Cry': Terrible Calls, Gallows Humour, and the Limits of 'Coping'

A powerful part of the literature on uniformed occupations, or others working in 'extreme' settings is its focus on swapping of stories. Storytelling is a central feature of many occupations and professions (Cousineau, 2016; Maitlis and Christensen, 2014; Maynard-Moody and Musheno, 2003), and in the rarefied world of 'extreme work' it often takes on some distinct features (Dingwall, 1977; Furness et al., 2021: 138–9; Granter et al., 2015; Tangherlini, 2000; van Hulst and Tsoukas, 2021). The stories are often unsettling and disturbing, sometimes mixing grisly stories of death, tragedy, and misfortune with jarring forms of black comedy. Many of the

---

[14] '"Toxic bullying" investigated at East of England ambulance trust', BBC News, 28 November 2019.
[15] 'Suicide prevention in the Ambulance Service', *Paramedic Insight*, 7(3): 18.

paramedics I encountered would share some of these stories, give their interpretation of the occupational meanings of these stories, and would comment on how swapping stories could sometimes be a powerful coping mechanism. By their nature, stories would only be shared if they were interesting. To get an airing, they had to be out of the ordinary and therefore memorable in some way. Some were memorable because they were weird, spectacular, or comic. Others were unforgettable because they were so horrible. The following is a fairly typical example of the strange dynamics I am exploring here, as ECSW Sally recalls a bizarre tale during a quiet moment on shift:

> We stop for a break, and sip takeout coffee in the street. Sally tells me that 'You remember some of the weird ones, the spooky ones. I remember once getting a call which came up as 'MAN STUCK IN WINDOW.' 'What the hell is that?', you think. And you get there and it is the strangest thing. There was this guy, in the weirdest position, sort of his legs hanging out from a window, and his clothes were so strange, it was like he was wearing rags. You thought 'Is this some sort of sick joke, or prank with a manikin or something?' But it was actually a person, someone maybe messing around or something, or was he locked out of his house or something? Anyway he was in fancy dress as some kind of monster and had climbed into a window, got halfway in, and slipped and trapped himself, sustaining catastrophic injuries. He was dead there, maybe all night. The paperboy found him and called it in. Poor little kid probably got the fright of his life.' It's a sick, sad story, but Sally tells it with a kind of shrug your shoulders, tragicomic, 'you know, shit happens,' type delivery.

To be precise, Sally wasn't really making a joke of this terrible incident. But, given the strange and uncompromising nature of a lot of ambulance service talk, she easily could have, and I wondered if this particular story has circulated ambulance stations in darker, more obviously comic formats. Some have argued that 'atrocity stories' (Dingwall, 1977) play important roles in the everyday social construction of a profession or occupation. In my time with ambulance crews I struggled to figure out any one 'function' of the gallows humour, or even if it could be said to have a function at all. Karen reflected in an interview on what she saw as the (rather limited) purposes of the 'sick humour'. In her view, joking about horror and tragedy can sometimes be cathartic, but mostly she saw it as part of the extensive and varied 'banter' of ambulance service culture:

Karen: There is still quite a lot of banter and most of it is well-meaning. There's a sick humour, but this is partly therapeutic, in that if you don't laugh you will cry. It temporarily lightens the mood, but it doesn't last.

Here we see gallows humour understood as part of an initial reaction to attending a nasty incident, helping the crews in the short term by sharing a difficult experience with others on station who have seen similar, or worse, and have similarly struggled. It helps to share the load, to disarm it, at least for a while. But Karen suggests that joking about the challenges of ambulance work has no lasting ameliorative effects. As such, it's not a very effective 'coping mechanism'. Instead, the swapping of stories is perhaps a broader phenomenon, more a part of informal cultural behaviours and not always clearly a 'coping' practice (Tangherlini, 1998, 2000). It is interesting how Karen wraps the sick humour into the broader culture of 'well-meaning banter'. Much like 'bottling it up', storytelling might work for the more minor forms of emotional strain, or it might work temporarily. But it seems increasingly clear that these kinds of informal practice are insufficient in dealing with the scale of the problem of mental ill health when a paramedic is severely challenged by acute trauma or when the slow build-up of strain leads to chronic mental illness.

This is not to say that story swapping is pointless or unimportant. Sometimes working on an ambulance can include a great deal of comedy that lifts the mood and, in a similar way to the work of Tangherlini, becomes the folklore and lifeblood of paramedic culture. I can illustrate this with an observation fieldnote that covers a range of these aspects. Included in this story are several comic tropes: management indifference to staff welfare, the oddness surrounding the enforced passivity of an incapacitated patient, and care for a fellow member of a culture that often processes trauma through camaraderie and 'pisstaking'. It was recounted by Sally and Chris near the end of their 12-hour shift, where there was little time left for Control to throw the crew another job. I could sense the crew relaxing and starting to think about getting off and going home:

On the drive back to the station the crew talk more about odd incidents. Sally mentions a near-disaster when a CTL fell out of a moving ambulance. But still it's recounted as if it were simply a funny story.
'Bless her, it was Emma who was driving and Dave was the CTL attending in the back. The patient was boarded and wasn't really able to speak. Suddenly Emma looked in her mirror and – no Dave! "What's happened, where is he?"!

The patient can't reply! She realizes the side door has failed and thinks Dave is dead in the road. It was on [a major A road]. She pulls over and calls Control: "Send everyone, my CTL's just fallen out the ambulance on the [A road]!" Can you imagine?!

Amazingly he was ok, he'd fallen out the side door and very luckily landed on a grass verge and was basically ok, with some fractures and bruising.'

Chris (a bit more serious in tone) 'Yeah, and you know what? It happened again 3 months later in Freeville. It was much worse, he was nearly killed. We're given repeated warnings about making sure that door is secured. I guess we're wise after the event.'

I wish they'd told me that before I got on the vehicle!

She shows me a photo on her phone of poor old Dave back at work. The crews had obviously made light of it by 'taking the piss'. He's sat in the chair in the back of an ambulance all taped up with yellow warning tape!

Sally: 'He's such a nice guy, Dave. He plays bass in a band, did you know that?'

Chris joins in now. 'Yeah, you know what his favourite song is?'

Me: 'No.'

Chris: *'Which Way Should I Jump?'*!

Sally: 'Haha, great! Who's his favourite band? The Doors?'

Lots of laughter. Chris is laughing too, now.

With the ground prepared and the mood lightened, Chris then shares a light-hearted story of his own. This one was typical of the range of 'Great British Public' type stories and anecdotes, which often play on the nature of the absence of medical knowledge of 'the punters', yet also reflects a certain type of relaxed interpersonal bearing and a kindness that comes with the paramedic role. Everyone is doing their best, but life is hard and death is always among us. Compared to the grim edge of the 'window monster' story, or the shock of the broken ambulance doors, I loved this anecdote's self-deprecating nature and its subtle humanity:

Chris: 'I went to a job the other day and it was literally next door to where I'd been a few days before to a man who had died. This second patient, they'd been to their neighbour's funeral literally either that same day, or the day before. As I'm there the widow from next door comes in and says: "Don't worry, Doris – he's the same lovely ambulanceman I had when my Jimmy died!"'

We can see that paramedics' comic interactions draw on both 'atrocity stories' (Dingwall, 1977) and 'self-mockery' (Cousineau, 2016) in the social construction of paramedic occupational culture. A crucial ingredient of many of the 'comic' stories was the strangeness of this line of work. Often it included themes around the inadequacies of the ambulance service equipment and management. The failing ambulance door is a classic example, but every ambulance station will have many others. Also very common was joking about the 'worried well', particularly by the more seasoned 'old school' paramedics who were often considered to be less forgiving of these calls. In David's account, the 'warped' ambulance service culture can produce crews who work together as a 'comedy double-act'. Stories of minor staff misbehaviour, revelling in cynicism and absurdity, were widely celebrated, meshing closely with the broader narrative of 'the moaning' discussed in Chapter 4. David describes banter as a way of escaping the 'vortex of doom and gloom'. It's 'the flip side of the moaning':

> David: There was a particularly warped culture in Metroville. It was very non-PC. [. . .] Sometimes you get a comedy double-act. There were these two guys who'd always be joking and sometimes the joke might be on patients. They told me once of being called out to a headache and they wrapped the BP lead around the patient's head, saying 'we're just looking to scan the inside of your head for anything wrong' and placed the blood sugar test strips up his nose, saying 'this is to check for the pressure in your nose, head, airways. . .!' The humour is often harsh, but funny - often very sharp and well timed. It's connected to the moaning, it's often a joke about management, about patients, other crews. To some extent the joking is the flip side of the moaning. Joking helps when we're just in that vortex of doom and gloom.

But other stories and reflections are much more serious than this, and often much nastier. Reasons for sharing the more grisly and upsetting stories were often said to be related to the need for 'decompressing'; a way to collectively digest and interpret some of the trauma. Sharing details of unpleasant and upsetting calls would show others on station that each of them had experienced something similar, serving to 'normalize' the experiences, to empathize, and to help defuse the story giver's anxiety, fear, guilt, and feelings of inadequacy and helplessness. If one paramedic, perhaps a junior one, had just attended a 'bad call', or had opened up to a colleague about one, then very often another paramedic would share stories of a similar experience they have had in order to try to lighten that load. Often there was no banter element here. It was much more about personal reflection of the nature of ambulance work, and of life, illness, and fate in general, things that are troublesome, philosophical and sad. Some of these mutually supportive stories and observations demonstrated how ambulance work can be 'character-building'. Experiencing this environment tends to invite a person to reflect on fundamental issues of life and death, such as the misfortune a patient experiences when everything possible was done by paramedics and other bystanders to help them, but that 'it was just their time to go'.

And yet, even here, there could be unsettling elements of comedy that sometimes sat in and around the horror and sadness. In an interview, Jamie makes an interesting allusion to the outrageous 'Springtime for Hitler' song from the 1967 movie *The Producers*:

> Jamie: I went to a terrible job on the [Motorway]. A suicidal man had leapt from a motorway bridge onto the carriageway. He landed head-first, and his head was just obliterated. I had some sleepless nights over this, just the image of it, the senselessness of it. And how ridiculous it looked. In a sense my mind has made a joke of it. Like Mel Brooks – Hitler as a joke. It's the only way to make sense of it.
>
> The way we discuss it on station does disarm it somewhat. People discussing lunch after they've been in contact with blood and vomit, shit, diarrhoea. You laugh at it. 'How long has he been that colour, do you think?'
>
> I went to a suicide in Danley, ten years ago. He had hung himself in his garage, using a hosepipe. It was summertime, steaming hot. You could tell from the decomposition that he'd been there a couple of days. Later my partner joked 'didn't he know there was a hosepipe ban on? Should have been nicked for that!'

As David Simon writes in *Homicide* '[i]n any other world, the comedy would be appalling' (Simon, 2006: 242). Jamie went on to make some further observations about sick humour within the context of 'how you rearrange your thoughts in your mind.' This part of his account tended to emphasize self-taught, private and personal techniques for dealing with trauma, rather than sharing it among crewmates or seeking formal or informal help from a line manager:

> Jamie: The gallows humour, it's how you rearrange your thoughts in your mind, like Tetris. Almost certainly there is PTSD. The issue is when it is chronic. You have to learn to rearrange how you deal with it, how you cope. I have read a bit about Neuro-Linguistic Programming. It makes you think of memories as pixels, as bits of Lego. You can re-arrange them. They are just emotions, visual, dark, light, you can blast them away, blast them into space. NLP was one way that has

influenced how I think about all this, and how I react to traumatic episodes. Let's say there was a horrible sound on a job. You think of a sound, you can change the pitch of it to Mickey Mouse, make it sound ridiculous, make it preposterous. It can't hurt me anymore.

Jamie's bold statement that 'it can't hurt me anymore' reminds me of Simon in Chapter 4 and his characterization of ambulance work as 'you and me against the world'. Both are suggestive of a traditional 'macho' approach, which favours self-control, independence, and individual competence; similar to the robustness of the 'armour plating' metaphor. Of course, all of this is necessary on some level if a person is to survive and function in this environment. But it is not sufficient if this environment is to be healthy. Excessive dependence on self-reliance and 'resilience' can mean that some will not seek out support and will not admit when they are becoming ill (Maunder et al., 2012). Self-learnt techniques such as 'mindfulness'—much less fads like NLP—are unlikely to be enough to provide effective solace for people exposed to truly terrible experiences. Whatever their statements to the contrary, I suspect that responders like Jamie are still being hurt by what they have endured.

'The Grief Mop'—Soaking up Poverty, Sadness, and Illness

Comparatively less dramatic issues can also be emotionally challenging. Many reflected on how ambulance work exposes paramedics and techs to a lot of sadness. Much like 'the bucket', one of the paramedics described this to me with the brilliant metaphor of 'the grief mop'. Calls that are not especially clinically taxing or interesting can nevertheless contain deep emotional challenges and can variously trigger stress reactions, philosophical reflection, and episodes of sympathy, learning, and peer support. Working out on the road brings frequent reminders of the extent of poverty, ill-health, and isolation across various regions and communities. Josh Seim's work in California resonates here, with his focus on the role of the ambulance as 'the frontline institution for governing urban suffering' (Seim, 2020: 12). NHS ambulances similarly respond to a tremendous volume of calls relating to general ill-health, poverty, substance addiction, and deprivation. Many of the calls I observed, in both urban and rural settings, possessed these features.

These calls are often not easy for ambulance crews to 'resolve', either at scene or by transportation to hospital. This means that the volume of these calls never decreases and that some patients develop into regular callers of 111 or 999. These 'social-type' calls are not very well regarded in the informal status hierarchy of meaningful or involved callouts, or what paramedics consider to be 'good jobs' or 'working jobs' (see Chapter 4). It can be hard for paramedics to 'offload' and 'decompress' about these calls, because their low acuity and high volume make them both (a) not interesting enough to warrant further discussion, and (b) not regarded as clinically or emotionally challenging and therefore not capable of upsetting or disturbing a competent responder. The following fieldnote provides some insights into this. Crew-room discussion circled around calls that were considered notable, and Emily's tiredness and exhaustion with a string of low-acuity calls does not merit much attention:

> I'm on an observation shift with solo paramedic Emily. We have just attended a call at a house whose occupants were heavy smokers. As we entered, we were hit by an overpowering smell of cigarette smoke. The house was filthy and the walls yellow-brown with tobacco and tar deposits. We weren't at the address more than about forty minutes, and the patient was handed off to another crew for transportation, but both of us now positively reek of cigarette smoke as we get back into the vehicle. Emily looks tired and dissatisfied.
> Emily: 'Urgh, it's gross. I stink.' She grabs the front of her shirt and sniffs it. I am also repelled by the stench of my clothes. I feel as if I've got a sheen of dirt all over my hands and face. Emily gives us both a squirt of a hand sanitizer gel called 'Scented Paradise', which I suppose helps a bit. 'The next job we go to, people will think we've been smoking. It makes us look unprofessional. I hate it when this happens.'
> Back at the station it is quite busy, with about three other people present. Emily checks the computer and talks to a youthful paramedic and a tall, experienced-looking ECSW. Another young female paramedic is coming in and out of the mess room and joins in some of the talk.
> Female ECSW: 'We've had a nice little easy day today.'
> Young woman: 'Yeah? We've been run ragged. Steve is still a trauma magnet, as ever. We've had four good jobs. Had a fit, then a TIA that seemed to be affecting one half of the patient's body. It was a weird one. I wanted to stay with the patient for longer all the way into hospital. Then another trauma, someone changing a tyre on his car and got hit by another car while crouching by the side of the road.'
> The discussion is interrupted as Emily again grasps part of her shirt and holds it

> up to her nose. The young male paramedic notices. 'Did you just smell
> yourself?' he asks. 'You stink!'

While Emily could easily have discussed the clinical pointlessness of the shift she was so far experiencing, and perhaps commented on the unpleasantness of reeking of cigarette smoke, instead she just doesn't bother to raise it.

These considerations speak to the contested purpose of ambulance work. What is discussed and digested on station goes to the heart of what the ambulance service is for, how its own expectations are understood from within, how it is (mis)understood by the public who call for it, and how it is portrayed in the media. I wrote up the following fieldnotes after observing a shift with CCP Becca. It was striking how quickly she moved the conversation from joking and criticizing TV personalities to describing 'real' trauma, in this case, drawing on an example from her professional experience of a call that did not require a great deal in terms of clinical intervention and risk, but that did carry extremely high emotional trauma for the patient and was very upsetting for the responder. As mentioned in Chapter 4, Becca was perhaps the most obvious 'trauma junkie' I came across, but here, after dismissing a TV personality's inauthentic portrayal of ambulance work, she goes on to show a side to her character that is much more emotionally engaged and empathetic:

> We talk about the portrayal of paramedics on TV. 'Ah, that bloody Angela whatsername?[16] She always cries! For fuck's sake, it's so fake and forced. I'm trying to think, when was the last time I cried?' She thinks for a bit, softens. The armour plating moves away just for a moment. 'I don't cry very often, but it does sometimes happen. The last time was a woman who'd been raped. She'd been badly beaten, her face was swollen and bruised. She was so distressed, I just felt so sorry for her.' I don't ask her any questions and I just stare though the windscreen at the road ahead. I'm left to wonder at the combination of emotions – sadness, sympathy, anger. I'm struggling to take this all in and I'm somewhat surprised by her totally candid approach to this. On the one hand the notion of her crying at anything is difficult to square as she comes over as so uncompromising and capable. But on the other hand, ambulance work is primarily about caring for others, and it would be difficult to imagine being a witness to this kind of suffering and not becoming profoundly upset.

[16] Becca is referring to the Sky One TV programme 'Emergency with Angela Griffin'.

As my notes suggest, gathering 'data' of this kind was a sometimes-awkward part of the research process. I was sometimes surprised by how open the paramedics (and occasionally their patients) could be about the brutalities of some of their experiences. I was aware that paramedics do not take kindly to inauthentic and inappropriate portrayals of their work. I didn't want to be in any way intrusive, insensitive or seen to be deliberately 'fishing' for the nastiest stories. In the following interview with Jamie, we see—similar to the above conversation with Becca—a paramedic criticizing the voyeuristic or inappropriate attitude of outsiders:

> Jamie: We get asked a lot 'what's the worst thing you've ever seen?' I remember I was getting my paramedic bags out of my car, and a teenage kid who lives next door sees the green equipment, the uniform, etc, and asked if I was a paramedic, and he asked me that: 'What's the worst thing you've ever seen?'
>
> I thought, 'why does he want to know this? What's the point? 'Do you really want to know? Really?' I said: 'The worst thing I've ever seen? Let's see. . . *The Sound of Music*!'

'What's the worst thing you've seen?' is an irritating question for paramedics, which they rarely answer directly (Seim, 2020: 68). And yet, in private, paramedics' swapping of tragic, disgusting and ridiculous stories is commonplace (Tangherlini, 1998). Ghastly incidents become part of the local folklore. Quite understandably, however, paramedics generally want to reproduce this folklore on their own terms. They are wary of being seen as uncaring of unprofessional if these stories (and especially those containing sick humour) escape the confines of the ambulance cab, the hospital corridor or the ambulance station mess room. The gallows humour around ghastly, bizarre, or intensely stupid incidents usually circulates in 'backstage' areas, by ambulance staff and for ambulance staff. It is part of what, in their mind, marks their occupation out as 'different' from those operating in 'normal' work conditions, whose work never brings them into contact with sexual violence, twisted car wrecks, patients having dangerous episodes of 'acute behavioural disturbance', or neglected, rotting bodies. Such story swapping is appropriate only when shared with others who inhabit this space and who understand its dynamics. Several paramedics mentioned that if employer interventions designed to help alleviate psychiatric distress are to be successful, then they would need to be conducted by other paramedics, or at least by people intimately familiar with the sadness, nastiness, and weirdness of emergency medical care.

Similar dynamics pertain to the general line management relationship. When paramedics feel that their line managers are not closely enough associated with 'the realities' of street work, then tensions and frustrations quickly arise. As the following section explores, line management relationships can be highly supportive, especially at frontline supervision level. But the broader management cultures are widely said to be inadequate, adding another potentially dangerous element into what is already a very challenging climate for professionalism, health, and wellbeing.

A Blame Culture

Much of the academic research on employee wellbeing emphasizes the central importance of supportive and trustworthy line managers, especially in intense and emotionally exhausting environments (Huo et al., 2022). I found the culture of line management in the ambulance service to be very varied in this regard. Some local line managers were excellent, and took a compassionate and supportive approach. Others were described as unapproachable, indifferent, and dismissive. Some appeared too busy to be able to offer much of anything to others. Paramedics would often refer to what they saw as a 'top heavy' management structure, with too few operational roadstaff available and 'too many managers' involved in duties that seemed to add little to the overall effort. Many spoke of managers as remote, 'old school', and obsessed with hitting targets, while also wasting time on trivial managerialist initiatives. Highly critical phrases would circulate among the roadstaff such as 'the fish rots from the head' and 'the beatings will continue until morale improves'. Gareth's comments below would be typical:

> Gareth: The biggest issue is management. They are not connected to staff. There is a lack of understanding. It's become very money-oriented, the patient side is in real decline. The next tier of management above us don't understand what we do. The things they are involved in, some of the management jobs and roles, I don't know what they are doing, either: 'Diversity for Tuesday... Quality because it's a Monday...' Managers should be able to do the job that roadstaff are doing. They should be able to carry out those duties, even if it's one shift per month. Keep their feet on the ground, see the difficulties we face, and the workload that we have to handle. We need to reduce the barriers between staff and managers.

It's important to note that ambulance line managers themselves were also stressed by various issues (Hyde et al., 2015). Responding to far fewer call-outs than regular roadstaff, managerscan often be especially sensitive to the issue of the skills drain when they do respond to emergency calls. Part of the perception among many paramedics is that some of the 'white shirts' have failed to keep up clinically with the newer cohorts. Many pointed to the ad hoc and traditional means whereby paramedics move into supervisor and managerial roles, with the promotion process seen as non-transparent, based on time served and on informal judgements. Well-respected and well-liked paramedics don't automatically make good managers. Many are reluctant managers who have to be dragged into the role. Perhaps the most common criticism made of managers was that they were unavailable or that 'you only see a manager when something has gone wrong'. This did not mean that managers would arrive after distressing incidents to provide emotional support. Quite the contrary. It implies that managers are remote from and indifferent towards staff wellbeing and that, rather than regarding difficult calls as times when a supervisor should be looking out for the welfare of their staff and thinking about opportunities for clinical learning and debriefing, managers instead go looking for roadstaff to blame following incoming complaints or poor results in the performance metrics.

Debriefs for ambulance staff after attending serious and distressing incidents such as suicides don't happen as often as they should. Paramedics have little or no specific training on how best to deal with certain types of often-distressing calls, particularly suicides (Nelson et al., 2020). Similar situations pertain to some of the most difficult, stressful, and haunting scenes that paramedics and techs will ever face, such as decisions to continue or stop resuscitation, especially as regards infant patients (Anderson et al., 2020; Reed et al., 2020). There is sometimes no debrief on 'a difficult job'. Paramedics attending such calls can carry with them the memories and scars for months or years to come. People can still be affected in some way by something that happened even 15 or 20 years ago, and they have no finality. There is a strong feeling among staff throughout the pre-hospital world that the cultures and practices of the established management system need to change. Ambulance services have a history of being a uniformed, 'boots on the ground' system, staffed by 'tough' individuals who are 'expected to handle anything'. This is a dated approach, insufficiently compassionate, professionally limiting and clinically not as advanced as it should be.

This old school approach was never perfect, but maybe it worked better when staff were less busy. On the road with Emily, she raised a range of criticisms that all stem from the 'operations' approach to ambulance services rather that its 'clinical' bearing. The pace of operations dominated everything that the Trust does, with paramedics getting 'chased' by management in its pursuit of response target times.

> The staff are not looked after. Staff welfare is not top of the priorities, let's say. The pressure remains the same but funding is being reduced. It is causing a lot of people to suffer with exhaustion. In the old days if you went to a nasty job, colleagues would understand, you might get a chance to go back to base station, do a restock, have a cup of tea. There would always be someone there you could talk to if you wanted. Whereas now you get chased to green up.

Karen spoke in considerable detail about the need for the ambulance service to develop a much more therapeutic management culture. Central to the complaints about management in general and certain managers in particular was a sense that the culture is not clinically advanced. Ambulance work on the ground has outgrown the limitations of the traditional command and control operational culture. Ambulance services were described as not being as clinically focused as they could be. Rather, the logics of operations, demand management, and target chasing dominate the agenda.

> Karen: I think many of us need ongoing therapy. Supervisors should be on the lookout for signs of stress among their road staff. Some staff do self-report their symptoms. Some are diagnosed with PTSD, and there are some moves towards taking this more seriously, actually providing proper time off and regular therapy. There is maybe now a better understanding of this developing at the centre. They are starting to adapt TRiM into the service (Trauma Risk Management), an idea taken from the Army. This is maybe ok for acute episodes of PTSD but if the symptoms are ongoing for more than one to two months and it becomes chronic, then maybe that person needs further intervention.
> I know that many suffer from what is effectively chronic PTSD. Some drink heavily. There is still a stigma attached to admitting you are struggling. We need to find a way to make these issues – these psychological problems – less of a stigma. We need to improve the way we handle this. We need people out there who are not just clinically trained but mentally ready.

There needs to be national guidance. It would help reduce sickness. We need a proven structure in place, a norm, so that going for treatment is not stigmatized. You can go for a session and see the help it can bring you. There's a bit of a frustration among staff in that outside counsellors are well-meaning and ok, but they lack the real experience of being on the road and what it entails. Sometimes they just don't have a clue what we face and what we do, and this can make opening up to someone hard to do. You need people we can relate to, they need to understand us. We need to somehow make it more real, maybe involve experienced paramedics themselves taking the counselling sessions.

Karen's account speaks to the need for authenticity, honesty and trust, within a more clinically advanced system where the emphasis is on learning and developing, not blaming and punishing. Examples of bullying and harassment are well known in English ambulance services (Lewis 2017; Manolchev and Lewis, 2021). This background reality surfaced in numerous ways. A few days after going out on an observation shift, I received an email from one of the paramedics I had been accompanying. Part of the email contained the following, which I found quite affecting:

I'm glad you found the third manning useful, and it was good to be able to give my views without any worry of reprisal.

Many staff feel unable to be fully honest in raising concerns, issues, or complaints with managers. The management needs to be substantially more sympathetic, open, and supportive. Karen spoken in detail about the blame culture of ambulance services and the intense fear of scrutiny, inspections, and suspensions that many staff carried with them:

Karen: There is a real sense of guilt among a lot of paramedics, when you felt, 'Did I do all I could, was my training up to scratch, did I do everything possible, did I miss something? Will management come down on me like a ton of bricks for missing something, will I face complaints?' Maybe it's better to not deal with it, hope it goes away? I think a massive issue with the amount of non-qualified staff is that it all falls on the paramedics' shoulders. They are the ones whose registrations are on the line if something goes wrong. Firefighters out in the field are all qualified. Only some of us in ambulance services are.

There's a lot of fear about losing registration, or getting suspended by HCPC. There have been situations when staff are too scared to open letters from their Trust, and had to ask their partners to open the envelope – they are so scared that the letter might be about some kind of investigation or suspension. That is the condition they are in.

There's been 3 suicides in [x] Trust in the last 5 years, all men. One was from our station. I had these posters and leaflets from Mind, the mental health charity, and I asked if I could place them around the station. I wasn't allowed to put the posters up, I was told 'let's wait until the dust has settled'. Management is so scared of having to admit anything was wrong on their side, they just are in denial about the scale of the problem. But they are also risk averse. They are scared of litigation and patient complaints, and areas like 111 and the EOC are totally risk averse. This means more workload, more callouts.

This isn't a new thing, either. Over the last ten years or so staff wellbeing has been really poor. What is relatively new is the people leaving. They are now leaving in droves.

I return to the issue of 'leaving in droves' in the following chapter. Like the 'five-year job' it was a common refrain and it shouldn't be taken on face value. Nor should the constant roadstaff complaints about managers and management. It is important to note that 'management' is not completely asleep at the wheel when it comes to staff mental health. It would be wrong to uncritically accept all of the dismal roadstaff narrative. Although they can be reluctant to seek support (Mars, 2020), ambulance staff do have access to it in the form of line management, union representatives, chaplains, TRiM counselling, and staff charities such as TASC.[17] The College of Paramedics has commissioned research and interventions into staff health and wellbeing. Many in England note that other countries have maybe progressed further on this, pointing to some important changes in Australia, such as the very comprehensive *People Behind 000* study, and at local levels where, for example, senior leadership of New South Wales Ambulance has apologized to staff it 'completely failed' due to an entrenched culture of bullying and toxicity.[18] More needs to be done to recognize precisely why staff don't access support. There are strong signs that the workplace culture is often not open, honest and supportive enough for staff to come forward, and that some believe there is little point in approaching managers when they are struggling because they suspect they will receive no substantial help.

[17] https://www.theasc.org.uk/
[18] 'NSW Ambulance boss apologises to paramedics it "completely failed" amid workplace bullying', ABC News, 26 June 2018.

Some aspects of the traditional managerial culture of ambulance services are slowly improving and changing. It is also important to sympathize with some of the issues that managers themselves face. Many ambulance managers work very close to the front line in hybrid clinical/managerial roles. They grapple with long hours, heavy responsibility, wide spans of control, intrusive scrutiny, and a sense that they 'can't win' no matter what they do (Hyde et al., 2015; Seim, 2020: 144–56). Their closeness to the line means they are knowledgeable about the intense demands of ambulance work. Other, non-operational management roles in ambulance trusts can also develop into 'extreme jobs' (Gascoigne et al., 2015). The responsibility to protect the public and staff (while sometimes putting the latter at risk) is a lot to take on. When things may have gone wrong and the public or uniformed staff have been harmed, managers can feel heavy burdens of guilt and regret. Simply being exposed to emergency conditions which include death, sickness, and loss can take its toll on mental health. Poor performance, perhaps in the form of continual missed performance targets, can also mean the ousting of top leaders (often in the glare of media publicity). It's not an easy environment for any person to inhabit, whether wearing the green or the white shirt.

I interviewed some more senior characters in several NHS trusts, including Consultant Paramedics and a Clinical Director. As expected, at these levels the picture portrayed is considerably less bleak. The comments below are from an interview with Steven, a Consultant Paramedic with over 20 years of experience of ambulance services from a student upwards. Steven wore the green uniform and projected that powerful hybrid bearing of clinician and manager. His self-image was that of the capable, confident manager, while possessing the powerful asset of clinical expertise that could win support from roadstaff, making him credible on station or at scenes:

Steven: I'm available on this radio all the time. Part of my role is clinical, I'm on a car responding on one shift per week. But also if there are certain complicated emergencies I can provide remote advice, I can help to talk through a scene with a paramedic, I can then call through on the phone to advise on a job. Sometimes if the scene is nearby I may go out to it on an ad hoc basis. In fact, all of us have the availability of the Clinical Support Hub, that can be there for many things from basic things like, say, you are a paramedic in West Banterdale and you are responding to a call that is out of your area, you can find out 'is there a falls team here?', for example – basic stuff – to more complicated things like getting more advanced clinical advice. There is a hierarchy from AP to me, to

Medical Director, 24/7 an escalation. So if it is something really difficult there is a shared decision-making. If the Medical Director himself is unsure then he will have access to his peers, you know, people he studied with.

In the old days you just made decisions yourself. A paramedic starting today has no concept of how much less support there used to be. In the old days managers were more remote and were simply dyed-in-the-wool managers, it was just general management, and they hadn't touched a patient in years, whereas now you have clinical support. This is what we mean by clinical leadership.

You still get this viewpoint from front-line responders 'I don't see a manager, I never see any managers.' But it's not true. I'm a manager, the APs are managers, there are CTLs on station and on the road, these guys are all managers to some extent. But they are not seen as such, an AP is just seen as another paramedic who is operational, but if you work with him or her in the field then they are supporting, they are helping.

Steven's viewpoint with its language of 'clinical leadership', was much more pro-management than those of most of the people I consulted in this book. Some of Steven's view was backed up by David, a relatively junior paramedic. Interestingly, David saw life on the road as 'mostly not stressful'. He provides some very important balance to the gloomy perspectives of the unsupported paramedic, potentially overwhelmed by risk and stress. He reminds us that paramedics always have the option of taking patients to hospital, and that they 'have backup' in the form of team leaders and HEMS:

David: But mostly the job is not stressful. Being on the road is enjoyable. Driving on blue lights is exciting, and you do have discretion, and improvising. I can't do a 9-5, when it's so predictable, you have none of that in an office-type job. In terms of patient care you do worry about getting things wrong or the patient rapidly deteriorating, but you've always got that option of taking to hospital. That does take some of the stress away. People say: 'You never know what you're gonna get' and that is kind of true although you can guarantee you'll get a granny fallen over, a diarrhoea and vomiting! But then they'll be the odd cheeky job when you really have to perform, something unexpected, something very serious. But you'll have backup, team leaders, HEMS.

We have returned to the contested concepts around professionalism. Steven's points about hybrid managers who are 'not seen as managers' is interesting, and speaks to the vital issue of who exactly should be a manager in this context and how they should act. Managers should respect front-line responders' discretion and autonomy, but they should also be available both in dynamic, real-time scenes and more broadly in the background to provide practical and clinical advice and support, and to encourage a culture of learning. There is a lot of truth in Steven's more optimistic narrative and at least some of this connects to the paramedics' professionalization agenda. Steven is a manager as well as a trusted, respected paramedic professional and a CoP member. Contrary to the claims of many paramedics, 'management' isn't always about extracting every last ounce of energy from roadstaff while turning a blind eye to their professional, clinical ambitions.

Conclusion

Part of being 'professional'—especially in its 'folk category' meaning—is the ability to focus, to cope, and to 'deal with' the challenges, stresses and strains of work in a given field. This is largely the responsibility of the individual. Professionals often work as employees in large organizations (Muzio et al., 2020). They are also usually part of a 'community of practice' (Gherardi, 2009), or 'occupational community' (van Maanen and Barley, 1984) which can take many forms. The employing organization and the wider professional community can provide ways for professionals to share their experiences, to provide mutual support and advice by formal and informal means. A person must clear a high bar of competence to be allowed into a profession, and a professional is (at least theoretically) equipped with the necessary qualifications, experiences, and tools. Professionals are given discretion and leeway, and are expected and trusted to get on with the job.

There are limits, however, to what a professional field or an organization can reasonably expect of a person. There is a balance to strike between the demands of the field and the assets that a professional possesses. Sadly, the tools and resources provided by the employer and by the community are often insufficient in meeting the demands of the field. The responsibilities for 'coping' rest heavily on the professional's shoulders and sometimes this burden is too much to bear. It is unreasonable to expect paramedics to bear this burden without due attention being paid to this imbalance in demands versus assets. The 'hero' narrative doesn't help. Paramedics aren't superhuman. They can't be expected to handle everything the field throws at them just by adopting the identity of 'professionalism', much less by having the 'hero' identity thrust upon them by the public or media (Granter et al., 2019: 286). Working in this field has real, hard requirements for the maintenance of paramedic psychological safety: supportive line management, more crew availability, better call prioritizing, more A&E capacity, less emphasis on targets, inspection and control, the removal of the blame culture, much more focus and time for clinical development, and the sharing of expertise. Professional burdens are loaded onto paramedics' shoulders. There is an unfair distribution of risk. The paramedic field—with its powerful social meaning, its rich symbolism, its legacy of robustness and toughness, and its growing clinical ambitions—is in danger of becoming a 'greedy institution' (Coser, 1974), a field that is intensely difficult to occupy, where one's individuality is potentially swallowed by the broader, greedy demands of fitting in, dealing with 'everything', personally embodying this demanding culture, and not raising complaints. Coser (1974: 17) writes:

> Membership in secret police organizations such as the FBI or the CIA, which require the loyalty of fiercely devoted members to do their 'dirty work,' or membership in warrior elites like the Teutonic Knights, the Templars, or the Green Berets might also be profitably studied from [that] angle of vision.

Ambulance staff are not 'warrior elites' and not all of their work is 'extreme' in its demands. But the pre-hospital field is an intense and challenging environment with a number of major risks and penalties that accompany the strong attractions and rewards of this professional calling. Based on the contents of this chapter, there are powerful reasons to doubt that this field provides sufficient resources and protections for its professional operators. The scales of demands versus assets are out of balance. Discourses of 'being professional' that place all the emphasis on individuals tipping their own personal scales back into balance by being 'resilient' and 'tough' aren't sufficient, and the evidence of this chapter shows how far we are from achieving equilibrium.

The crisis in ambulance worker ill-health is hardly unknown. There is considerable public and some political recognition of how unforgiving emergency work can be, and that much more needs to be done to protect the welfare of emergency responders. Blue-light mental health has become an important topic of discussion among charities, professional associations, trade unions, and employers. The inadequacy of healthcare provision for the hundreds of emergency staff whose health was affected by the 9/11 terrorist attacks was a major political controversy in the United States. More recently the UK Parliament passed new legislation that increases the potential criminal penalties for assaults on emergency workers.[19] In keeping with developments across society, occupational taboos about mental health issues seem to be lifting, with emergency staff becoming more confident in expressing their anxieties over burnout and PTSD, sometimes publicly via social media, or by employer news output. Emergency services employers have improved their capacities for recognizing and treating depression, anxiety, and PTSD, but there are signs that the provision is still far from sufficient.[20]

Some of the most 'extreme' elements of ambulance work are largely unavoidable, but much more could be done to lessen their impacts. Addressing some of the field's other ugly and unpleasant attributes is also entirely possible—the blame culture, the lack of in-service training and debriefing, and the generally unsupportive management conditions. Much more can be done to improve the working culture, reduce operational strain, and address the crisis in employee wellbeing. All of these changes need to happen if the ambulance world is to more closely resemble a field where professionals can truly practise their craft in a more sustainable setting.

My characterization of the paramedic field in England places it broadly in the category of a 'new profession' (Evetts, 2011). It is relatively young, having a restricted formal knowledge base that historically was 'extended' out to them from the more established medical profession. NHS paramedics are employees in large bureaucratic organizations. They are, to an extent, controlled by managers, as well as being regulated by external agencies. They have relatively little control over the structures of their field, but considerable frontline autonomy in their dealings with patients. This chapter has explored the personal hazards that ambulance work poses to paramedics and the various 'coping mechanisms' that occupational members have constructed around themselves. Taken together, the multiple risks and threats to wellbeing and livelihood represent powerful challenges faced by paramedicine in its ongoing mission to develop, augment, and embody professionalism. The social meaning, moral value, social popularity,

---

[19] 'Prison sentences doubled for people who attack emergency services workers under new law', *The Independent*, 13 September 2018.
[20] 'Fire staff on long-term mental-health leave up by 30%', BBC News, 17 September 2017.

and altruistic bearing of paramedic work lend themselves very well to classic notions of professionalism. But hard field realities impose considerable limits and barriers on the prospects for this profession to deepen and flourish. I further explore how the paramedics understand the complex reality of their profession in this harsh context in the next, and final, empirical chapter.

6
Street-Level Professionals

'The semi-professionals' efforts to change themselves, more fully to live up to the claim floated, generate a major source of tensions because there are several powerful societal limitations on the extent to which these occupations can be fully professionalized.'

(Etzioni, 1969: vii)

A professionalization project is an ambitious mission. Securing a legally protected title and a licence to practise are major achievements. Success for a profession, however, is rarely permanent and irrevocable. Professional status can decline. The erosion of status is increasingly the experience of many professions under conditions of neoliberal competition, technological change, managerialism, and radical scepticism about experts' claims to knowledge and altruism. Operator autonomy is curtailed by managerial control, by client and market demand, and by audits, checklists, and performance management systems. Professions battle not only to achieve status, but to retain it. This scenario is clearly recognizable at a structural and abstract analytical level, in which 'professions' are understood as self-contained groups possessing a degree of discretion and agency.

But to think of professions in this collective, structural, and abstract way is not very enlightening if we don't also consider the individuals who construct, enact, and embody professional roles, behaviours, and identities. The social influence, authority, and importance of a profession is not represented only by what exists legally on paper for that group, such as their levels of qualification or the lists of technical or clinical procedures they are licensed to carry out. 'Professionalism' also has intangible connotations about behavioural and attitudinal expectations (Freidson, 2001; Muzio et al., 2020). Acting 'professionally' (or 'unprofessionally') carries powerful folk meanings. The holding of a professional office affords certain benefits (perhaps secured collectively through a professionalization project), but it also entails significant individual responsibilities. All professionals—perhaps especially those wearing a uniform and placed in a public-facing role—must always consider the responsibilities of that office.

The Paramedic at Work. Leo McCann, Oxford University Press.
© Leo McCann (2022). DOI: 10.1093/oso/9780198816362.003.0006

For a paramedic this is obviously most true when working 'on scene' with patients and bystanders. But the enactment of professionalism also occurs in other settings and modes and at deeper, more personal, levels often in 'backstage' settings, such as during interactions with other clinicians and occupations.

Being a professional paramedic entails the expending of effort towards developing and sustaining appropriate forms of behaviour or bearing. This extends beyond interpersonal behaviour and into certain aspects of the private and personal makeup of the individual. At a minimum, there is the obvious mandated requirement to keep up to date with changing clinical guidelines and to observe practical and ethical requirements so that they are—in the language of clinical professionals and AHPs—'fit to practise'. In addition, there are more subtle and informal cultural and behavioural demands. These are manifest, for example, in the ways in which a paramedic will develop private strategies to 'manage' the demands, strains, and frustrations of their work. There are tremendous spoken and unspoken expectations for a professional responder to be 'resilient'. They are expected to respond competently to emergencies meaning they need to secure adequate mental and emotional space for themselves in which they can work effectively and safely. Society's expectations for effective, timely, compassionate, ethical, and personalized care are growing, even for the 'semi-professions' who are increasingly being assessed according to the high expectations of 'professional' clinicians without necessarily receiving the status, autonomy, and reward that those groups have traditionally received. Professionalism is something that many occupations pursue, but the enactment of professionalism is not something taken on lightly and is not all about conferring rewards on the members of a professional community. It also confers personal duties and responsibilities. It carries risks and imposes costs.

Professionalism can also be thought of as *as a social condition*, as an activity; a form of interaction or service that an occupation, an individual, or a workplace aspires to take part in and construct on a regular basis—ideally as the default basis. I describe this as *professionalism in process*. When matters come together well in the enactment of professional work, then the professional has done his or her job effectively, the client has received a professional service, and 'the public' has been 'served'. In our case, these are the intangible and much sought-after incidents where a paramedic has 'made a difference' or where he or she has attended 'a good job' or 'a working job' (see Chapter 4). In the complex and multifaceted world of pre-hospital medicine, the achievement of professionalism as process is highly dependent on context. Sometimes it can be achieved, sometimes it cannot. Approached in this way, professionalism is not only a form of status or bearing that a person or occupation holds. Professionalism in process is something broader, less tangible, and less permanent—something that exists in the moment when a service is rendered effectively (Korczynski, 2007; Lopez, 2010). Professionalism is 'an accomplishment' to use a fashionable sociological phrase. Its achievement depends not only on the skills, expertise, judgement, actions, and bearing of the professional expert, but also on the organization employing that expert. It is also contingent on the behaviour of the client, in our case the patient. When patients present with minor ailments or show a lack of courtesy and respect to the responder, and where an ambulance responder is constantly dispatched to 'nothing' and 'rubbish' calls, then professionalism in process is unlikely to arise.

This chapter will explore paramedics' understanding of professionalism at both the abstract, structural level of the occupation, and also at the more tangible, more personal levels of the individual self and at the operating context where paramedic work takes place—at Lipsky's legendary 'street-level'. Building on several themes already explored in the book, this chapter aims to identify the essence of paramedic professionalism at work. Its overall aim is to bring into sharper focus the idea of paramedics as 'street-level professionals', a notion drawing inspiration from both Michael Lipsky's 'street-level bureaucracy' (2010/1980) and from a much less well-known idea—Donald Metz's characterization of ambulance workers as 'blue-collar professionals' (1981). My aim in conceptualizing paramedics as 'street-level professionals' is to highlight the complexities and struggles of enacting and embodying professionalism in an often-difficult environment, while also considering what the status of 'professionalism' really brings to an occupation and its members. As we have seen, the pre-hospital field can be decidedly *inhospitable* to professionalism as process. But that doesn't mean that professionalism cannot be enacted there or that mid-level professions' pursuit of professional status is somehow a hopeless quest.

The chapter opens at the macro level, exploring paramedics' understandings of the core of the profession, something occasionally referred to in the businesslike language of its 'unique selling point'. These issues are becoming increasingly disputed by changes associated with the professionalization project. In an interesting sign of change, university degree qualifications now equip paramedics with the ability to leave ambulance trusts but to remain working elsewhere in the health economy under a paramedic registration. It will move on to explore the chronic mismeasuring of ambulance work, and the ways in which the prospects for 'good' paramedic work are so often blocked by the allocation to ambulance crews of clinically inappropriate duties by risk-averse call prioritization and by a public that fails to understand what constitutes acceptable use of the 999 or even the 111 service. 'Management' and the EOC are supposed to act as a filter that screens out inappropriate calls, but paramedics have little faith it that process. In the eyes of paramedics, management's ineffective filtering of calls and its insensitivity to clinical development severely hamper the prospects for paramedics to 'properly' exercise their craft. The chapter also explores the importance of training and education. Once qualified and employed, paramedics experience few further formal opportunities for clinical development. But ample scope exists for street-level professionals to engage in less structured forms of learning in the field. I will show that informal modes of learning, advice, and mutual support goes on regularly between crews. Having explored these structural elements, the chapter shifts to explore the more personal and individual elements of professionalism in this environment, discussing the ways in which paramedics construct, enact, and police forms of conduct deemed 'acceptable' within the profession. Critically, this involves enacting humble, stoical, and self-deprecating behaviours, rejecting the 'hero' narrative, and understanding professionalism in terms of a clinical and patient-centred approach to work and conduct. The chapter ends by exploring what good paramedic work looks like—a form of professionalism in process which emphasizes practical and compassionate forms of care, and where paramedics try to find ways to keep the many and varied frustrations of street work from interrupting and disturbing their goals. Throughout, I aim to keep the broader conception of 'street-level professionalism' in view, as the book moves towards its conclusion.

'That Ability to Respond': Emergencies as the Lifeblood of Paramedic Professionalism

Paramedics are routinely frustrated by what they regard as their misuse by management and by the general public. We have seen how they complain extensively about 'rubbish', 'pointless', and 'nothing' calls. The informal value hierarchy of ambulance callouts places complex trauma and CPR calls at the summit of respect and places drunks, elderly falls, childhood fevers, and cut fingers at the base. And yet, for all this, there is still a lack of clarity about what the 'core business' of a paramedic is. This confusion seems to stem from the increasing breadth and complexity of ambulance patient demand, and the growing trend for ambulance services to provide 'community paramedic' or unplanned primary care models. The introduction of specialist paramedic roles also leads to some unintended consequences. Discourses of patient-centred healthcare (Berwick, 2009) also disrupt the profession's traditional understanding of appropriate and worthy callouts, by (theoretically) shifting the power dynamic from the clinician to the patient.

A very useful source of information on this subject was Sarah, a highly experienced paramedic who has regularly worked on secondment to the College of Paramedics. We discussed in a long interview how the development of more advanced clinical career pathways has created new uncertainties about what might be considered a 'core' paramedic job description. Perhaps counter-intuitively, a focus on enhancing clinical care at the pinnacle of the traditional ambulance value hierarchy (such as developing dedicated CCP or RAT crews to focus on trauma and CPR) potentially risks placing the provision of Advanced Life Support beyond the competence of 'ordinary' paramedic crews. Sarah claimed this was illogical. It is 'just wrong'. She was adamant that traumas and arrests are the lifeblood of paramedic work. At play here is not just a clinical logic but also a cultural one; paramedics are acclimated towards out-of-hospital emergencies. When on scene at a potentially life-threatening emergency they are at their highest value to society. Life-threatening emergencies are what ambulance crews are mentally and operationally prepared for. It is what they seek out. It is the essence of the paramedic calling, 'the basis of the job':

Sarah:[1] We keep having this discussion, 'what is our unique selling point'? It's weird, surely the USP should include ALS? It's the basis of the job, it's what you are trained for and want to be involved in. I can see what happens when you train up this one area, you leave another area behind in comparison, but the idea that an ambulance crew can't handle a trauma or a CPR is just wrong. Everyone needs to be competent at that. Also, there is the simple fact that an advanced crew just won't be available to get to every emergency requiring ALS.

---

[1] All names of people, places, and organizations are pseudonyms. For details on the research process involved with the writing of this book, please see the Appendix.

That all paramedic crews need to be able to handle these kinds of call is undeniable. But there is also a logic behind establishing advanced specialist clinical pathways to provide career headroom for those in the service who want to build their clinical competence rather than being funnelled into line management. The development of such roles also speaks to the dynamic whereby clinical occupations typically aspire to accumulate increasingly advanced clinical competencies. Even if everyone agrees that trauma and resuscitation are absolutely core parts of the paramedic role, there remains ethical, operational, and financial confusions about how advanced or specialized their trauma intervention should be. How often would the most advanced emergency skills be used, and what are the risks associated with training clinicians in invasive skills that they rarely get a chance to practise?

And then there is the 'community paramedic' or 'unplanned primary care' element to consider. As we have seen, these typically low-acuity and psychiatric-based calls are taking up an increasingly large share of the paramedic workload. Paramedics are well qualified to treat such patients. And the type of attentive interaction that can be fostered by a direct call-out to a patient makes the paramedic very well placed to deliver care that is truly 'person-centred'. But why is the NHS sending them to these patients in an ambulance on an emergency callout basis? Rather than being delivered via a 111 or 999 call and a blue-light run, surely it makes much more sense for these skills to be made available to patients in a community setting, such as a GP surgery or walk-in centre? For these reasons, many paramedics—especially those employed as PPs or similar—are leaving NHS ambulance trusts to continue their work as clinicians under paramedic HCPC registration but in different settings with different employers. Their skills are in high demand and in short supply.

I asked Sarah about this development. On one level it seems a very powerful and even necessary element of the professionalization project. Professionals shouldn't be dependent on their employer for their expert status. Nor should they be restrained or captured by employers. Paramedics who leave the ambulance service to work in community settings could be seen as having successfully outgrown the limitations of their ambulance trust employers. But there are reasons to have reservations about this. The concept of a paramedic who doesn't work on an ambulance, a response car or a helicopter seems strange and unsettling. If a paramedic chooses to work in a more sedate clinical setting, then they have broken from the informal status hierarchy of calls and from the 'response to emergencies' that

Sarah described as 'the basis of the job'. Understandably, Sarah explained the logic and value of this development in terms of the relative positioning of paramedics within broader NHS clinical hierarchies. Even here, however, she circled back to the ambulance, and specifically to the notion of 'emergency':

> Leo: It's an odd thing to some extent, a paramedic who doesn't work on an ambulance, like a police officer who doesn't work in a police force?
> Sarah: Yeah, when you compare it to blue light services that doesn't make sense, I see what you mean. But with the broader NHS family then it does make sense, especially if you compare it to nursing. The training and background might have been paramedic science, but this is more than enough for a lot of nursing type work, care work, clinical work. Working in a walk-in centre, a GP's surgery, that is ok, I don't have a problem with that. What makes no sense for me is new graduates of the degree in paramedic science not ever working on an ambulance, just graduating then going straight to some kind of physician-assistant type work, never going to emergencies, that seems wrong. For me a lot of the USP is responding to emergencies, being able to do it, being ready for them. Even if this is outside NHS ambulance trusts, that's fine, you know, it might be a standby, private ambulance for medical care at large public events, that's a different setting from ambulance trusts, but is still an emergency way of working. Even working on a cruise ship or an oil rig or something like that. I don't like the idea of being a paramedic if you've never done it, you've never responded to emergencies. What defines a paramedic is that ability to respond. Working in a GP surgery, these aren't emergencies. I've been to 999 calls to GP surgeries and often the practice nurse, often her background doesn't prepare her properly for this kind of patient episode. I went to a young girl having a fit at a GPs, and when I arrive the nurse is giving CPR, but she wasn't in respiratory arrest! I took a look at her and she's breathing! I've had situations where the person with a nursing background has been more of a hindrance than a help; life-threatening emergency is not what they are trained to see and do.

What defines a paramedic is that ability to respond. For all its clinical weaknesses and frustrations, that operational, emergency basis remains essential to the paramedic identity. The idea of the paramedic in uniform responding to calls out in the field is such a powerful idea—one that seems considerably more attractive to 'edgework'-minded paramedics than the alternative of sitting in a clinic or office. In Sarah's account they could even be working in some isolated or exotic place such as such as an oil rig or a cruise ship. The paramedics are inexorably drawn to the calling of the outdoors, to mobility, to the practice and improvisational work of *responding*. Sadly, it seems, in the world of degree-trained professions, working out on the street seems outside of the usual cultural expectations. Traditional, perhaps class-infused, ideas of 'professional' work tend not to conjure up images of a person wearing a uniform and steel-capped boots, sent from pillar to post according to the commands and measurements of a demand-led, 'operational' regime. Paramedics are experts in a form of work that is not readily valued or understood in terms of society's traditional, middle-class notions of what constitutes 'professional' work. Many types of clinician can climb the professional ladder without leaving the confines of the scholarly, clinical world and exposing themselves to the rawness of street-level work.[2] An ambulance paramedic doesn't have that luxury of choice. But, to a degree, they like it that way. There is a certain romance surrounding the stoical 'responder' who looks forward to seeing whatever bundle of random events the next call might bring. It is this scenario that stimulates many of the cultural tropes surrounding the 'realities' of ambulance working life; the stoical sense of completing endless shifts without 'Control' really understanding the nature of the work, the self-deprecating banter, the 'old school' industrial relations of 'roadstaff' versus 'management'. The pre-hospital clinical environment has a cultural life all of its own. Ambulance workers often complain about their comparatively lower status in relation to other clinicians, but you sometimes feel they wouldn't have it any other way.

Helen, an Education and Training manager at an NHS ambulance trust, spoke of the 'odd hybrid' of ambulance culture and the risks of 'snobbery' that can come with professionalization. A focus on building clinical competence can lead to the neglect of some of the traditional elements of ambulance work and of people in support roles such as EMTs. She also mentioned how professionalization gives paramedics a way out of the service that they didn't have before. Speaking as a manager, this is clearly a problem for employers, in that they 'lose paramedics as quickly as we bring them in':

Helen: It's an odd culture in the ambulance service, its part NHS, part emergency services. In some sense it is similar to the police, in dealing with the worst

---

[2] Should they wish to, trauma-specialist doctors can work out on the road, on a helicopter, or even in a warzone on a voluntary or secondment basis. They usually receive huge clinical respect from others in the pre-hospital field for doing so.

aspects of things, getting used to bad outcomes, the uniform, the boots, the vehicles. But it's an odd hybrid. The philosophy of the paramedics is to help people, it doesn't have that law-and-order element that the police have. It's a caring element and that is much more like core NHS. It also has that feel of a family, the good and the bad of it. [. . .] Some people have been here thirty years. It also has the bureaucracy and the management issues and problems that are also typical NHS. There are loads of people here who were PTS and moved up the ranks. More than you might realise at first. This means there's a snobbery around professionalization. On one hand the medical directorate says 'we must do that', we need the ambulance service to be more clinically-focused, with higher qualifications. There's a non-stated element that the newer paramedics with degrees are better than the traditional guys. And that is a bit dangerous. [. . .] We have to value these people, all that experience, that has come up through the ranks from PTS. One issue is that the training that seems to be the most valued is the advanced clinical stuff.

The tensions around all this is that the professionalization has led to valuing people from outside coming in. That is good, it is necessary. But there is the issue of potential neglect of what we already have. We're now starting to look more closely at how we grow our own and pay more attention to our career pipeline. Now the workforce is very complex, there is a combination of lots of different ranges of job. [. . .] We have entry trade, Band 3 right through to 3,4,5,6, all of whom are roadstaff, operational.

By her account, the ambulance service recognizes the importance of upgrading clinical skills. But without the cultural ability to handle pre-hospital interactions, working on the streets, going into people's homes, and 'getting used to bad outcomes'—like a police officer—then this expanded clinical aptitude cannot be put to good use. She astutely sees the cultural contradictions of ambulance expertise. White-collar, academic *clinical professionalism* exists alongside the more blue-collar, street-level *emergency professionalism*. The practice of a competent paramedic should exhibit both of these elements, and a well-run ambulance service needs to understand, value and reward them both.

NHS trusts are struggling with this. We have seen how poorly these employers score in staff satisfaction surveys. It is understandable that many paramedics distance themselves from the employing trusts, often in a mental sense and, for some of them, also in terms of formal employment status. Like nurses and other AHPs, increasing numbers of paramedics are moving themselves onto 'bank' contracts whereby they can select fewer (or perhaps more) shifts over certain time periods, and where they can select which station to 'book on' to a shift, and possibly also who to crew with if coordinating with another 'bank' paramedic. This is one way for paramedics to retain their professional registration but to work on other areas, often to pursue higher degrees, or a rotational role elsewhere in the NHS. It is easy to envisage further developments in this area. Volunteer community first responders, or ambulance trust managers with a paramedic registration can already 'book on' to the EOC emergency response system using a smartphone app and respond to calls in their own private cars.[3] Market competition is ever-present on the fringes of this profession and it doesn't take a huge leap to imagine a much broader 'gig economy' or platform economy for responders, operating on a similar model to taxis and ride-hailing apps. Professions need to develop, adapt, and grow, but they also need an employment domain that supports and protects them. If publicly funded NHS trusts can't do this well enough, then the market will find other places for paramedics, with all the risks that would entail. As we shall see in the following section, there are many areas where the existing structures and demands of NHS ambulance trusts inhibit the fuller flourishing of paramedic professionalism, potentially driving people away.

111 Is a Joke: Unreasonable Patient Demand as an Inhibitor of Professional Growth

The NHS, as a service 'free at the point of use', has long struggled with overwhelming demand (Hyde et al., 2016; Taylor, 2013). How to ration this demand, to triage and prioritize it, to assess the risks involved in calculating how long a patient will have to wait for assessment and treatment, is a desperate challenge. Heavy demand, and clinically inappropriate demand, are two of the most serious problems that paramedics face. This problem was widely said to be one of the most serious and obvious obstacles hindering the development of a more clinically advanced and professionally rewarding ambulance service.

General Practitioners, A&E units, and the 999 emergency service represent the three main entry points for patients into the NHS system. Each is

---

[3] For an illustration, see the images in this Tweet: https://twitter.com/searchy_boy/status/1337382520929853442?s=20

chronically overloaded. In March 1998, NHS Direct, a 24-hour nurse-led telephone advice service, was set up to provide an alternative pathway, assessing patients' conditions over the telephone aiming to provide advice for patients to administer self-care and hence reduce unnecessary NHS activity. An immediate problem faced by NHS Direct was clinical risk assessment. Its response was to be famously risk averse. When in doubt, calls would often be passed to the ambulance service, with crews jokingly referring to the service as 'NHS Redirect'. NHS Direct was replaced by the 111 service in 2014, a non-emergency telephone system designed to better coordinate the various services that patients can access if required, including the development—alongside ambulance Trusts—of 'Hear and Treat' protocols whereby experienced clinicians located in ambulance EOCs can make better decisions about whether to advise a patient over the phone or to mobilize an ambulance response. Often the service works well. But, like any managerial or organization innovation, it will have unintended outcomes.

Frustration with 111 is regularly expressed by ambulance crews. They typically describe calls that escalate from 111 as 'kids with colds', 'worried parents'. They claim that the 111 call handlers give out bad advice which increases paramedics' workloads. Complaints about 111 were part of the broader narrative about unnecessary 999 calls from patients being compounded by risk averse call prioritizing. The assumption was that the 111 service, and those running the EOC 'Hear and Treat' and 'See and Treat' protocols, were out of touch, thereby creating more work—and more clinically unnecessary work—for street-level responders. The processes by which the EOC dispatches an ambulance to a call are a mystery to ambulance crews, and call prioritizing systems such as AMPDS (Advanced Medical Priority Dispatch System) were widely described as 'a joke'.

The targets culture is often described as a lurking presence behind the questionable prioritizing. Today's response times are still framed by the legacy of ORCON—a set of response time standards calculated in 1974 by the systems analysis methodologies of an organization known as Operational Research Consultancy.[4] Paramedics barely existed in the Seventies. Back then ambulance services were clinically very limited, 'scoop and run' services. The ORCON standards were based on the assumptions of a much older model where ambulance staff are essentially major trauma

---

[4] See Matthew Westhorpe's excellent blog post on the 'Cult of ORCON', 9 October 2012: https://westhorpe.net/2012/10/the-cult-of-orcon/

and cardiac emergency responders, and when callout volume was far lower than today. Paramedics would suggest that the pressure to hit the response time targets (including automating some of the dispatch process so that it would task the nearest available vehicle) meant that decisions were taken in a rush, without due care as to whether dispatching a crew is sensible for that call. Closely bound up with paramedics' critique of Control were the claims that management is poorly informed, lacks clinical skills, or seems disconnected and uninterested in the clinical details of patient care. 'What matters' (Corman, 2017b) seems to be something other than paramedics' assessment of patient need, such as performance targets, time spent on scene, time taken to mobilize, time of arrival on scene, or any of the range of data artefacts that constitute ambulance services' audit culture. Demand pressures are reminiscent of the Soviet economy where factory workers would 'storm' in order to meet sudden and arbitrary deadlines, breaking people and equipment in doing so. In the background are the rather grim-sounding 'REAP levels': the four colour-coded levels of the Resource Escalation Action Plan. Here, Clinical Care Paramedic Rob describes an ambulance service 'burning the family furniture' in order to quickly respond to 'the dross':

> Rob: It's all hands on deck. We're burning the family furniture to keep the fire going. It's getting ridiculous, the secondary waits: anything from 30 minutes to 3 hours. Then there is still all the dross. Government wants more out of us in terms of performance and they won't budge on the targets and quality indicators. The need to send a resource immediately – it has a stranglehold on everything. It's bonkers. Short term targets dominate, clinically they don't make sense. If we send the wrong resource and the patient dies but we arrived within 8 minutes then it's ok, but if the right resource is sent but arrives after 8 minutes it's recorded as a fail.

Anyone in the ambulance service will have heard this refrain many times. But there were variations on this theme. Others would be reluctant, for example, to describe non-emergency calls as 'dross' and would make interesting observations about 111-type patients needing some kind of response. If the ambulance service is partly there to provide care, advice, and reassurance, then such calls are appropriate and part of the job. It's often simply unclear what a patient needs, or what duties and expectations rest on which organizations and occupations to provide it. This problem is particularly intense when so many calls to 111 or 999 come from persons experiencing various forms of suffering that interlock and overlap: physical and mental illness, poverty, isolation, inadequate or zero social support, and lacking capacity for self-care. Inadequate and overstretched services elsewhere (housing, social care, mental health provision) mean that chronic problems of low clinical acuity (such as social and psychological issues) morph into emergencies now requiring urgent intervention (McCann and Granter, 2019). It is actually not easy to judge what is, and what is not, the 'appropriate' use of 111, 999, and A&E units (Lowe and Abbuhl, 2001). It's similarly difficult to judge when a crew was dispatched 'appropriately' or not. Sarah was again highly articulate about this problem, discussing the rise of 'social work type' calls:

> Sarah: The ['reality'] TV programmes go on about crank callers and timewasters. They do exist but they are very rare. What is constant, is people who are sick, but don't require an emergency ambulance. That is a huge amount of our work. [. . .]
> Leo: You see a paramedic venting on station and he or she will say 'I've been to nothing but crap today'.
> Sarah: Yeah, well the definition of 'crap' seems to get wider all the time. I seem to attract lots of social work type calls, not a lot of trauma. When I was young, there was more trauma, it was a higher percentage of your callouts because there were just fewer callouts. So, when people did call it was often more likely to be a real emergency. But if I was starting now, as a new paramedic, I'd be worried if I was never getting a trauma job. It would prey on my mind; 'will I remember what to do, will I be able to cope with it? Will I be sharp, will I switch on?' So, for another reason some of the paramedics, maybe some of the younger ones, are thinking of leaving because there isn't enough trauma – you aren't getting the practice, the experience you need to handle them. I've noticed that there seems to be less talking about jobs, less swapping of experience. Partly it's because you are off station all the time, partly also people don't know each other as much. There are lots of people on station you've not even met, or haven't been able to get to know. I can go on station and meet no-one I know.

It is a difficult balance. There is a need to treat all patients with dignity, but not to completely pivot the ambulance service to unplanned primary care to the extent that paramedics don't experience 'enough trauma' to be able to 'switch on' at 'real emergencies' involving threat to life, or that all of its capacity is taken up with low acuity, and often largely irresolvable, 'social work type' calls.

What is the ambulance service for? What constitutes a 'real' emergency to which a paramedic can properly exercise 'that ability to respond'? Those with more advanced clinical training (such as Becca and Rob) would complain frequently about being sent to calls that did not require a Critical Care Paramedic crew, and indeed probably not an ambulance response of any kind. And it also tended to be the 'old school' type responders who would be less sympathetic to these 'clearly not an emergency' type calls. Some of the younger, degree-trained paramedics tended to be more accepting that every call has to be taken seriously and the person calling needs care of some kind. There is often no-one else available to do it. And low acuity, social care calls are now explicitly part of the evolution of the ambulance service from an emergency service towards a deliverer of unplanned primary care. If a paramedic can visit a patient, perform an assessment, provide advice and reassurance, and perhaps even prescribe, then this is another person kept out of hospital or their GP's surgery. It's also another vehicle not tied up in a queue outside A&E. Both of these are good outcomes. The adaptation of the service and the expectations surrounding it is another element contributing to the slow decline of the 'trauma junkie' persona. Conversely, however, it was mentioned to me several times (such as by Sarah above) that the 'new breed', degree-trained paramedics were getting disaffected with working in ambulance services because there were experiencing too many social care calls and too few calls that really drew on their clinical education.

Triaging calls is a complex matter, especially when resources are so stretched. Paramedics and techs expressed frustration about how unclear the goals and priorities of the organization had become. Some of the old-school crews would ask 'why can't we just go back to scoop and run?' Innovations would come in only to be later dropped. For example, at one time the RRV seemed to herald the arrival of a new 'front-loaded' ambulance model. But many trusts have recently cut back on their usage of RRVs, restricting them mostly to city centres. Roadstaff perceived few practical improvements from changes made in 2017 to the national prioritizing and call response standards based on the Ambulance Response Programme.[5] Chronic problems about call volumes and insufficient vehicles and crews were met with a kind of tired, resigned reflection. More vehicles and crews would surely help, but it would also mean more of them sitting in a queue outside crammed-full hospitals. The ambulance operator culture was one

[5] For details on the Ambulance Response Programme of 2017, see: https://www.england.nhs.uk/urgent-emergency-care/improving-ambulance-services/arp/

in which there is little point in speculating too much about what management is doing right or wrong, or how it could all be better organized. Instead, they learn to just 'see what's out there', to react as best as they can to each 'job' they are assigned to, using as much clinical autonomy and practical common sense as possible to try to figure out the best approach for each patient.

A crucial blockage towards a more substantial form of professional autonomy, therefore, is the simple fact that paramedics can't police the size and nature of the workload that comes their way. Many in the pre-hospital community have urged for EOCs to be given more time to triage the calls in more detail rather than to 'fire off' the nearest vehicle as soon as possible to meet a time-based target. ARP has been a welcome step in that direction, but paramedics of all backgrounds still regularly complain about being overwhelmed with inappropriate, low-acuity calls.

> Rob: New paramedics are saying 'this is not what I signed up for.' The 111 jobs
> are a real drain. We're losing people, the new graduates are not as loyal, and
> they also lack the life experience you need for this job.

Paramedics aren't the only ones struggling with this problem. Professionals working in any large organization will often confront forms of work they view as inappropriate. Academics find themselves teaching huge numbers of high fee-paying students, some of whom don't meet the standards hoped. Doctors are struggling with terrible backlogs of unmanageable patient care. Schoolteachers talk of the need for 'professionally acceptable' workloads. Conflict regarding overwhelming and inappropriate demand represents a disconnect between managerial and professional expectations of what the ambulance service is for. As the next section will show, there is a strong sense among middle and senior managers that the 'operational reality' of massive patient places unavoidable limitations on ambulance trusts' ability to support and realize the growing clinical aspirations of the paramedics they employ.

How 'the Old Ambulance Service Way' Restricts Paramedic Learning and Development

Paramedics don't get to influence the shape of the window through which patients come to them. This is controlled by EOCs and managers, who, for their part, also despair at the scale and nature of calls they have to process

(Hyde et al., 2016). In such a scenario, managers often resort to a line of argument that 'this is the reality we face, and you either deal with it or leave'. Karen reflected on the lack of sympathy that managers often have for paramedics' complaints, something entrenched in the rather dismal, pessimistic culture of 'old-school ambulancemen':

> Karen: It's like that back to the 1970s feel, of 'ambulancemen', a man in a van, which tended to have kind of a bullying culture. A lot of the managers are old-school ambulancemen. They've never known anything else, only ambulances. They've been around ages and they reinforce each other's behaviour. They have come through together and they don't learn or are not interested in others. They act as a clique. It's who you know. The culture is not open, not transparent. They don't inform you of what is going on. People feel unsettled by this. They don't know what is coming next and it is hard for them to adapt and prepare for it. [. . .] Unions also don't represent us well. They often don't keep us informed, either. Some feel unions are in management's pocket.

The type of paramedic interested above all in trauma and critical care would tend to argue that the skills profile and scope of practice for paramedics as still too constrained, with NHS trusts focused solely on target times and unwilling to encourage the expansion of clinical scope, autonomy and versatility. Rob was a good example of this viewpoint, recalling the decision for paramedic intubation to be withdrawn by a decision of JRCALC back in 2008:

> Rob: Look at the row over endotracheal intubation – anaesthetists have narrowed their focus so much that they're defensive, they are worried about losing out to us now. It was a bit like that at times during the in-hospital training – I'm present at Resus and some of them are like: 'Who is he, what does he want to learn? And why?' I'll be polite, but I'll tell them why. Consultants are too expensive. And do they really want to be out on the road with us at 4am? We have a paramedic-led pre-hospital environment and some of these skills need to be available out on the road.

Paramedic Jamie similarly scorned the managers, 'othering' them as 'the white shirts'. He found them guilty of holding back the paramedic profession, and he yearned for greater clinical discretion and autonomy. His critique of the managers suggested years of frustration and conflict. He clearly resented being 'managed' by people who he felt had little clinical appreciation:

> Jamie: The management, they lack insight, they don't understand why you are upset with them. 'The answer is within', as a kung fu master would say! Why do we feel devalued? It's the same people, the same values. They bring consultants in to do a staff survey, but it will be the same thing again and again. We don't want gold-plated cannulas, or Gucci boots. We just want some clinical discretion for ourselves. In reality we are entrusted with a lot of responsibility, but we don't have the respect to go with it. The culture needs to be more Socratic, it is not asking any questions of itself.
>
> The white shirts just aren't respected clinically. A lot of them effectively have less clinical training than an HCA in a hospital. Their skills aren't up to date. They are in the role because of their familiarity on station. The white shirts – they work together, they cover things up sometimes. They are very remote and they don't recognize your skills. They should know if you know your shit or if you don't. And if you don't, you should be stuffed, but it doesn't happen. They don't know the standards of their workforce. If they worked more closely out with us, they would know, but they don't.
>
> If they worked out with us, it would soon become clear how little some of them know. They would be embarrassed and exposed. It is a recurring pattern. They are exposed as inadequate when out in the field, there is a tacit acknowledgement of this. Sometimes you just see it in a raised eyebrow.

Some paramedics were demoralized by the feeling that the ambulance service is 'carrying' weaker staff, many of whom have nonetheless found their way up into managerial ranks. 'Old school' managers were portrayed as not fully grasping the changing clinical demands and requirements of paramedic work. Because ambulance work is misunderstood, misallocated, and poorly measured, there was a sense in which the NHS trusts did not formally understand the 'quality' and 'safety' of some of the people operating on the streets or managing ambulance stations. Some were said to lack clinical confidence and would attempt to obscure this from others. Support and training are required to address this problem, but is not forthcoming in an organization that lacks time and resources and suffers under a blame culture.

In complaining about management and parts of the roadstaff workforce, the paramedics' grievances are interesting. Their concerns are not like the nostalgic criticisms of established professions such as medicine or law who dream of turning back the clock to a time when there was more autonomy, more time and little or no managerialist scrutiny. Rather, the comparative youth of the paramedic profession meant that it had yet to build in the professional boundaries that would have excluded some of the weaker entrants coming in. Clinical standards have risen to the point that many of the 'old school' wouldn't qualify for entry now, and yet paramedics suffer under their weak, rank-based, and outdated forms of command and control.

Frustration was also expressed by paramedics having to work alongside less qualified responders. I can illuminate this with a vignette from a shift on the road with Paramedic Practitioner Katie. We were sent to back up a private-sector ambulance crew for a 'man collapsed' at a railway station. Although Katie worked freely and easily with the private sector crew, she was clearly unimpressed with their clinical capacity and felt dragged into a low-acuity job that any ambulance crew worthy of the name should have been able to handle without assistance. As she greened up for the next available call, she described the callout just completed as 'a babysitting job':

It's 1720. The radio bleeps and we've got another call, this time to a railway station for a '50 YOM collapsed'. We hurry to the vehicle and climb on. Katie: 'I hope it's not a resus and they haven't told me.' It's now pretty much afternoon rush hour traffic and the roads are congested. It takes quite a while to get there, probably about ten minutes. The sound of the siren seems to echo off the sides of other vehicles when we are close: lorries and buses in particular. There is often so little room, and Katie is battling to get through tight spaces and round sometimes almost blind corners. It's quite tricky for some of the cars to get out of our way at times.

Upon arrival we see an ambulance already on scene: Katie immediately groans: 'Ah, it's a private crew.' I can see it's not an NHS vehicle; the large ambulance is of different design from the NHS Mercedes; taller, less boxy at the rear and with a private sector corporate logo.

Rail station staff had moved the patient to a little staff office. Inside this cramped room we find two very tall and burly women in a different shade of green uniform from the one Katie is wearing. But the interaction is pretty much as with another NHS crew. The patient is connected to a small portable ECG and is sat on a chair. He looks exhausted but is in quite good spirits and again this doesn't appear life-threatening. Katie reads the ECG readout and asks the patient and the railway staff about what has happened and whether the patient has any underlying conditions. The patient seems ok. He is able to stand and walk, so Katie disconnects the ECG leads and we all head back to the parked ambulances.

Katie takes the patient up into the private crew's ambulance to continue the patient assessment and she asks me to bring the large ECG from our vehicle. The patient lies on the trolley and Katie connects up the 12-lead. The private crew ask about the positioning of the electrodes and Katie explains what she is doing, informally passing on her knowledge. One of the crewmembers says: 'It's sometimes such a relief to have these lovely PPs'.

We get back into our vehicle. 'Ok, Leo', says Katie. 'That was what we call a babysitting job!'

The use of private crews is clearly a response to the challenges of the targets regime, but their use can be self-defeating if their clinical abilities are so low that NHS staff are sent to 'babysit' them. Contracting private sector crews is the latest in a long line of management innovations that have largely failed to resolve the patient demand crisis. The government, NHS trusts, management consultants, and various clinical professional bodies regularly conduct investigations, reports, audits, and redesigns of the technical side of ambulance dispatch, deployment, and performance management. There have been attempts to generate 'better data' such as the rise of clinical outcome measures. But roadstaff see little real impact of these innovations and modifications. There is never a change of clinical culture. Instead we see the persistence of management just 'throwing more stuff' at the problem:

Rob: Outcome measures haven't changed this culture at all. We've not got the right number of clinically focused people in management. People don't want to do the management work. The wrong ones are getting promoted up. Even those at the top, the managers, they also get bogged down, they lose sight of the bigger picture, they require some direction. They need guidance, they need things simplified.

[. . .]

We're having a bad time, we're missing targets, the REAP levels, and we're dealing with it in the old ambulance management way: just throw more stuff at it. They can't learn, they are unable to learn.

Part of 'the old ambulance management way' is the rank structure, something that seems increasingly out of step with the aim of clinically developing the ambulance service. Although paramedics are normally quite sympathetic to police (typically less so to fire and rescue), they often argued that the culture, image and status of the ambulance service should be moving closer to the 'NHS family' and its promise of clinical innovation, and away from its 'blue light' emergency services roots. Clinical learning and development seem to promise a clearer route to 'professionalism'. But that goal was said to be short-circuited by the traditional roots of the ambulance service, dominated by an 'operations' logic of command and control and hidebound by performance regimes and the fear of failure.

These criticisms are very instructive. But it would be wrong to assert that the life of the operational paramedic involves no learning at all. The following section explores the nature of experiential learning in NHS ambulance services, and the important, informal role it plays in the social construction of paramedic professionalism.

Professional Knowledge, Learning, and Bearing

Learning, education, and knowledge are critical aspects of a profession. This is true not only in terms of the formal credentials required to become a qualified and licensed practitioner, but also in the often more informal, everyday elements of developing, reflecting, and learning throughout a career (Muzio et al., 2020: 57). Education, knowledge, skills, and formal qualifications are clearly an important element of 'professional' clinical practice. But their relative and absolute value is contested in certain instructive ways.

Firstly, and relating to the points raised by Helen above, many in the ambulance world are somewhat ambivalent about the move towards an all-graduate intake for new HCPC registration. There are complaints about the time, costs, and practical value of university study. The shift to all-degree entry has largely removed the traditional pathway for those in other ambulance roles to 'work their way up' to a paramedic qualification as an ambulance trust employee in service. Today, if a technician wanted to qualify as a paramedic, then they would most likely have to start on a paramedic science degree among cohorts of typically much younger students, or try to find an opportunity of a workplace-supported degree apprenticeship. Some complain that the degree privileges book-based, abstract learning, and pays insufficient attention to the cultural complexities involved in being a competent paramedic on the road. The range of people qualified as a paramedic is varied—recall the 'huge spectrum of ability' mentioned in Chapter 3. How is 'ability' to be measured? It is often intangible. The achievement of clinical qualifications does not automatically make a person an effective, competent, well-respected paramedic who is compassionate towards patients and acts appropriately on and off scene.

Secondly, qualified and operational paramedics have few formal opportunities for training and learning while employed because of massive callout demand and the organizational needs of keeping rosters full. Additional training is not a trust priority. Professionalization and HCPC registration have contributed to a situation where trusts can more easily justify shifting the responsibility to be clinically up to date onto the shoulders of individual registrants.

Thirdly, however, there is a more optimistic reading of the phenomena of paramedic learning. Examples of knowledge and skills generation, and teaching and instruction are continually visible, often in an informal and everyday sense. I would often witness paramedics and techs advising each other, supporting one another, and providing reassurance. Much of the core learning and practising about how to be effective paramedic is experiential, intangible and unmeasured. Staff would converse with one another on station, in the ambulance cab, and in hospital corridors about recent calls they have attended, sometimes asking each other questions about which treatments were the most appropriate, and seeking reassurance that what they did was correct.

As a completely naïve observer from outside the ranks, I came to learn this myself a few times on observation shifts. For example, back at a standby post not long after a CPR call that sadly was unable to resuscitate a patient, a highly experienced technician spoke philosophically to me about the difficult feelings associated with being unable to save a patient from death. He recalled responding to a CPR call at a campsite where a patient suffered a cardiac arrest and couldn't be saved despite the presence of two paramedics, one tech, and two off-duty doctors and a firefighter who happened to be staying at the camp. 'He was just gonna go.' At other times following a 'bad outcome' for a patient (typically an unrecoverable CPR) there would sometimes be something called a 'hot debrief' of everyone present at the scene. Before everyone heads back to their vehicles to 'green up', the most experienced paramedic on scene would do a quick check around of everyone present to ask if anyone has any questions: 'Do you think that went as well as it could? Did anything happen that you didn't understand? Is anyone concerned by anything they saw?'

Discussions and behaviours like these entail the development and enaction of role models in the field. This is a subject of great discussion in the ambulance world, given all the issues surrounding wellbeing, burnout, and poor standards of management. Rather than a clinical, supportive and learning culture, ambulance trusts are often portrayed as having a culture of bullying, blame, and fear of failure. Newly qualified paramedics can struggle with the responsibility of being on the HCPC register. University education and expanding the scope of clinical knowledge are vital for upgrading the status of the paramedic profession, but they run the risk of somehow degrading the value of informal and experiential learning. And where the most sought-after clinical knowledge tends to slot into the high-risk, high-acuity value hierarchy, then expertise about how best to handle the uncertain 'social-type' calls can sometimes be neglected. This further contributes to the problematic situation where callouts to unplanned primary care episodes are often not considered valuable uses of paramedics' time. Practice educators need to enact constructive and supportive attitudes because their behaviours and opinions will 'rub off' on students.

And, of course, like all occupational groups, paramedics do at times make errors, act dishonestly, cause harm and attempt to deny and conceal harmful and incompetent behaviour. Personally unpleasant or clinically incompetent behaviour can lead to employers issuing gross misconduct investigations and for paramedics to be reported to the HCPC and potentially struck off the register following fitness to practise hearings. Browsing through the publicly available judgements of HCPC hearings can be unpleasant. One will read allegations of paramedics (and other AHPs) engaged in all sorts of questionable actions, from what might be considered relatively minor infractions involving 'banter' or criticism of patients and the NHS, to very serious issues involving clinical errors, failing to report errors, insults of a sexual, ethnic, or classed nature, inappropriate touching of colleagues, use of excessive physical force on a patient, and even theft of clinically controlled substances. Alongside cases where a paramedic is clearly at fault, sadly there are also frequent examples of registered paramedics being deemed unfit to practise due to sickness, many of which are likely to be related to mental health.

Discussions of personal professional standards are, however, inseparable from the context in which a professional is employed, especially given the issue of appropriate role models. One of the most senior managers I spoke to was Linda, who was clearly concerned about what she saw as a lack of learning from within the occupation, a persistence of an 'old school', 'lifer' mentality, and poor role models. But she believed this is changing with professionalization, particularly as people are increasingly filtering into and out of ambulance trust employers, changing the culture via the renewal of personnel:

Linda: The overall employment culture, the industrial relations has been extremely bad for a long time. It's been a culture of grievances, suspensions, blame. There's been many race discrimination complaints and cases. [Ambulance Trust] is a single profession, a monoculture. If you look at a Foundation Trust Hospital, it is made up of hundreds of professions, occupations, specialisms. Here it is overwhelmingly ambulance trust people, whether paramedics, technicians, other roles in the EOC, etc, and they haven't known anything outside of ambulance trusts, which are a peculiar culture. There are professional rows in ambulance services that I haven't seen anywhere near as much in other NHS organizations, such as: 'she's a nurse, I'm not listening to her'. There is a major problem with 'Not Invented Here Syndrome'.

Another thing that is unique to ambulance services is that no-one moves. There hasn't been a culture of rotation of people in and out of trusts into different kinds of organization. That is also unhealthy. People don't learn. The trusts are so large and wide, that people take up another role elsewhere in the trust, rather than go to another employer. They've been here for 20, 30 years.

There should be more chances for paramedics to move out of an ambulance trust, experience something else for a few years, put their abilities to good use in another setting, learn a few other ways of working. If they really miss life on the road then they will eventually come back. Again, in Foundation Trusts there is more of a tradition of placements where a nurse, for example, would be placed on different wards, then A&E, which ambulances lack. Now there are changes starting to happen where they can be rotated through various settings: prisons, Primary Care, Palliative Care. It is slowly changing, slowly getting better on this.

If you look at who rises to the top in the old context then it means you are getting a narrow group of people coming through. Chief Execs are all from the same community. They are all highly institutionalized. The power relations, the possibilities for abuse are all there. They are nearly all men. They have operated in a climate where they have the power to determine your career. This filters right down to lower levels, in some cases in extreme ways that are unreasonable. A lot of it operates on gossip, tradition, rumours about who is who, how they behave, maybe some whiff of scandal or failure in their past. Such as 'we all know about Dave, he's not safe, I'd not work on a vehicle with him, and he won't get promoted under me', because of some long-standing issue. An issue that might not even have been Dave's fault. I learnt of a situation where an ECA was trying to move to an ECT job, he'd been trying for 15 years! Eventually it did come about, in the sense of 'well, maybe it's now his time'.

Given the endless criticism from roadstaff about the 'out of touch' management, Linda's account of the institutional inhibitions to a culture of learning seemed to correspond very closely to the portrayal provided by ambulance crews. Some managers were clearly aware of the problem and articulated some powerful ideas about how the culture is changing, if slowly and incompletely.

There is also the important reality that the mobile nature of ambulance work itself can provide unexpected occasions for clinical learning. Fieldwork sometimes allowed me to observe examples of learning from unusual experiences, and from other healthcare professionals in other settings. The following fieldnote describes a cardiac case in which the ambulance crews' observations suggests some confusion over the diagnoses, but also demonstrates how all the parties involved in a patient's care contributed to shared learning and a good outcome. This call provided me with a fascinating glimpse into the growing integration of the paramedic into hospital care, firstly with a doctor at an A&E department, then with a cardiac technician who allowed the paramedic crew to stay in the hospital catheritization suite control room and experience his world for a while. The ambulance crew felt welcomed, not isolated, belittled or misunderstood. While the integration wasn't seamless, the ambulance crew's actions were nevertheless a major part of the practising of 'gold standard' (Campeau, 2016) clinical care by the NHS as a whole.

We arrive at a call to 'Chest pains and DIB'. An RV is already on scene, and Chris recognises the call sign as Lisa, a PP from his station. The patient is a 51YOF, reporting a pain score of 5. She is lying on the floor connected to the PP's Lifepak. A growing mass of ECG printouts creep across the floor. Chris and Lisa study them. There is an ashtray stuffed with cigarette ends.
'I smoke a lot', the patient admits.
She looks pretty unwell and seems uncomfortable. The crews are both not altogether sure what is wrong. 'We can give her a bit of IV para [intravenous paracetamol] in the truck.'
'Ow, I don't like ambulances,' the patient says. 'Oh shit, don't take me.'
Chris: 'We have to, we need to find out what is wrong with you. We think it

might not be a heart attack, but you will need more observations in the hospital.' The ECG suggests a borderline STEMI but both Chris and Lisa think this isn't enough evidence to take her straight to the hospital that has the 'cath lab' [catheterization suite]. Lisa has tapped the patient details into the ECG and has sent the readout to the hospital. She gets on the phone to a clinical desk which advises that the patient is taken to A&E and assessed from there. The patient is taken to the ambulance in the carry chair and Sally drives to the hospital on blue lights.

After a short ride to hospital, the A&E doctor takes the handover from Chris. Chris explains how he called ahead and was unsure of STEMI, so was advised to go here. He makes it clear he is happy to re-transport to the cath lab if needed. The A&E doctor reappears shortly and explains that Chris has to transport the patient to the cath lab at a nearby hospital immediately, saying there is a small but clear STEMI on the latest readout.

For a moment I think Chris is a little defensive in that he reminds the doctor that 'the clinical desk sent me here', but the doctor is ok about it all and agrees that the output is complex and ambivalent, and that they do change. This, most recent, reading is clearer. So we're heading back out with this patient. No-one is rushed. It all proceeds at a calm pace and the situation is explained to the patient. She asks: 'Chris, will you take me?' He seemed pleased that the patient asked him that she and used his name. 'Yes, it's us, we'll take you', he replies.

We drive over to the other hospital. The patient is pretty talkative and seems in reasonably good spirits. Chris continues to monitor her using his 12 lead ECG. 'It is not certain how they'll treat you', he explains, 'but it's likely you'll be taken right to the cath lab for the angioplasty.'

'When will this happen?', asks the patient.

'Straight away if they agree it's needed.'

The patient is clearly concerned, but also impressed with it all. 'Isn't this brilliant, eh? The treatment they can give you. . .'

It doesn't take us long to arrive and she is indeed wheeled straight through a labyrinth of corridors towards the PPCI unit which is signposted clearly with regular arrows and signs along the way. I notice lots of corporate sloganeering around the hospital: 'outstanding patient care', 'world class leadership', etc.

The cardiac catheterization laboratory or 'cath lab' looks like a hi-tech operating theatre. It contains several complex and sturdy-looking machines. Several devices mounted on booms hang from the ceiling. There is a door to a little control room that sits behind large panel windows. We head through that door, past a rack from which hang an assortment of lead-lined aprons. In the control room is a youngish looking technician looking over five VDU screens of information. He has a relaxed, friendly and welcoming manner. The patient is now the hospital's responsibility, yet there's a kind of extended transition point in which the ambulance crew is still on the fringes of the patient episode, observing for a while, and taking this opportunity for learning and reflecting. Chris explains to me: 'Years ago we used to thrombolize but this was often a bit hit and miss. This procedure is much more focused and targeted on the blockage specifically.'

There is a total of around eight people working on this patient. The patient is quickly given a local anaesthetic and they get to work. There's a consultant cardiologist, two nurses, plus radiographers and technicians. They have a radio link to the tech in the control room.

One of the screens shows live X-ray video, it is in greyscale and shows a lot of activity in the area where the catheter is being inserted. The screen is fascinating to watch. The catheter appears clearly as a solid white line in a sea of grey and cloudy human tissue. The cord heads down a blood vessel and seems to loop around in a complex arc up through the arm and into the chest cavity, curling into what looks like a hook. I assume this is the balloon and stent on the end of the catheter.

The whole area is pulsing. The vessels fill up with blood, become a brighter colour on the screen, then pulse out again. Vessels fan out all over the place. They look like inverted trees. The technician explains that the catheter also allows the spread of dye into the vessels so the team can see how the blood is moving and where the occlusions are.

The patient calls out in pain a couple of times and the tech says something rather unkind along the likes of 'she's being a bit of a baby.'

He's providing a kind of running commentary to the ambulance crew. 'Here, that's the dye going in. [. . .] Looks like he's having trouble getting to the blockage'.

The cardiologist is a little condescending:

'Now, Mrs Brown, we're attempting to get to the blockage but it is taking us a little longer than we'd like. There are some significant blockages in your arteries, what with all the smoking and what have you.'

The control room tech shakes his head. 'So, she's had an MI but she's continued smoking. Ridiculous.'

The tech turns to Chris: 'What was the call to balloon time? When did you get the call? About 10am? It's now noon!'

Chris: 'Yeah, but the ECG changed. I think the chest pains probably started much earlier and she was a little bit untruthful at times.'

We've had a while there, about 20 minutes and it's now time to move on. The

tech says thanks, was very welcoming, and asks the ambulance crew how their
shift has been.

Several interesting aspects are visible in this episode. It was a rare example
where an ambulance crew got to see the care of the patient extended into
the hospital setting rather than being something that may or may not happen once a patient had been transferred. Chris and Sally worked together
as a broader NHS team which, as a whole, showed sympathy, efficiency,
and care to the patient (alongside some private criticism). The patient
responded well to Chris' honest explanations and careful, attentive demeanour. 'Call to balloon time' on this call would have been just within the
guideline 150 minutes that represents the upper time limit used in clinical
audits, and the call was slowed by the ambiguous early ECGs taken in the
field. But overall the whole episode appeared to be an excellent outcome,
with the patient actually praising the system to Chris in the back of the
ambulance. The scene presents us with another demonstration of the inadequacies of performance management. Patients are not always truthful
about their current condition or their behaviour. Ambulance crews drive
long distances between calls, negotiate heavy traffic, and diagnoses can
change during a call. Given the giant scale of NHS operations, one could
argue that call to balloon time and other call response standards are very
rough, macro indicators that do provide useful data about care quality and
safety at an aggregate level. But they are not much use in measuring the
nature, process, and outcomes of individual cases. The numbers on this
particular job won't look especially encouraging. But the patient got what
she needed and 'professionalism in practice' certainly took place.

This call gave the paramedic crew a relatively rare opportunity to take
part in the enactment of a clinical 'gold standard'. Although paramedics are
not at the forefront of developing and shaping 'gold standards', I found that
they often shared their own interpretations and opinions about what constitutes best practices for patients. Several expressed concerns that things
are done in the ambulance service that might no longer be clinically up to
date. Specialist equipment, drugs, and protocols aren't always made available to them. Paramedics would complain that it often falls on their own
shoulders to turn up to CPD events on their own time, often to training and information events organized by the CoP and by volunteers. NHS
trusts seemed to take little responsibility for this. Some claimed that clinical
guidelines were unclear, such as the procedures for the use of spinal boards
and immobilizing patients. I heard skepticism from some about whether
nebulizers work very well in treating asthma attacks. Some worried about

what they saw as very limited guidance around the dreadful prospect of attending cases of Sudden Infant Death Syndrome. While observing a shift with degree-trained paramedic Emily, she reflects on what she saw as some quite severe clinical limitations in the ambulance service culture. Familiar with the clinical language, she spoke of the need for 'an evidence base':

> There is a lot of strange behaviours, unpleasant behaviours. People who talk in terms of 'this is what we've always done, it's just the way it is.' The culture is entrenched. We need more of an evidence base about what we do if we're ever to be taken seriously among other professions. [. . .] We have lots of guidelines and protocols but there is a lack of real training time to go with them. You do get that while at university studying for the degree. You have to learn them and you can learn them properly, but outside of that, once you're employed in an ambulance trust, if you are not studying on a course you have no hope – you can't learn and develop your skills. This won't change unless there's a change in how the service is funded. Extra training time is not gonna happen. The Trust has its core targets and these dominate everything.

David, also degree-trained, reflected on these issues repeatedly. For David, clinical limitations and risks are not only the result of an entrenched ambulance service culture, but also are structural problems associated with the nature of operating in an out-of-hospital environment. This environment can be 'debilitating':

> David: I studied at University of Midtown, and I believe this is one of the best degrees around. It is a good balance of road and classroom. You get a lot of whining that there is too much classroom, and not enough practical experience being gained or passed on. We don't really understand the underlying physiology. That is getting to be increasingly a problem as we deal more and more with urgent care. [. . .] We have no real access to patient info, no patient history. A GP will have a stack of records, we often have little or nothing to go on beyond what the patients or bystanders tell us. It can be debilitating. You often want to know what happened to the patient. You can sometimes find out later if you know the nurse at the hospital. But often you have no way of knowing if your care was helpful to the patient, or how the whole episode or treatment panned out for that individual. This is a key issue with our progression as a profession. We are not party to their broader care and treatment. We're given directions. Such as COPD nebuliser [Chronic Obstructive Pulmonary Disorder] for six minutes

to raise O2 levels, but we don't see the effects of this down the line. Sometimes you are gambling a little bit and you don't really see the outcome, it is disjointed.

In a debilitating environment, there is only so much you can do. Professionalism in process can be difficult to achieve. But the fact that it is hard to achieve should not prevent the professional paramedic from attempting to achieve it. Even when it can't happen, there is a degree of solace to be had in the stoical perseverance of doing all that one can. As we will see in the next section, this is an important element of paramedic professionalism. Even when the calls are clinically boring, when the patient misbehaves, and when all around there is exhaustion and cynicism, paramedics can still pursue their humane, self-sacrificing aims as 'street-level professionals'. They can harbour a disposition that combines the difficult experiences of the 'street-level' with more general ideas of altruism associated with a professional duty to put the needs of the public and the profession ahead of the needs of one's own.

'You Don't Get Hero Biscuits': Informal Norms of 'Authentic' Professionalism

Consider the following Tweet, posted by a Critical Care Paramedic and former RAF medic:

> Fellow pre-hospital service providers. . . PLEASE stop tweeting jobs! If you need your ego massaging then I suggest you need to evaluate your career. We get paid to do this privileged job that's reward enough. You don't get hero biscuits for doing your job! @MacTheCCP 24 September 2018[6]

His comments highlight several relevant aspects of paramedic culture and professionalism. Many emergency services and NHS personnel use Twitter and other social media platforms to connect with others, share information, and to support campaigns. Employers such as police forces, fire and rescue services, and ambulance trusts will also issue marketing-type content via various channels, such as a good news story about a recent incident or new initiative, or advice to would-be 999 callers. Paramedics are often critical of these marketing type Tweets, seeing them as symptomatic

---

[6] https://twitter.com/macTheCcp/status/1044305353834463233?s=20

of the inauthentic and detached culture of Trust management. Perhaps worse in their eyes are Tweets from fellow paramedics that are written in a similarly boastful style, especially ones that come across as too much like pro-employer marketing, or that have the potential to breach patient confidentiality. The notion of 'hero biscuits' is an amusing rejoinder, that also hits back at the 'hero' portrayal of emergency responders that the public is typically fed from 'reality' TV. Paramedics will regularly mention their disdain for TV programmes featuring ambulance crews. One programme with the title *Total Emergency* was laughingly referred to as *Total Bollocks*. If a paramedic seeks to publicly endorse the hero notion, or to ascribe that notion to themselves, then they are acting selfishly, in a way that reflects badly on the profession as a whole.

These concerns speak to intangible, unspoken norms around behaving professionally, about embodying the paramedic role in ways that members of the profession at large deem to be appropriate. Sounding too friendly to management and the employer is frowned upon. Pretty much any form of boasting is wrong. Fishing for recognition from outside the profession is particularly tasteless. Something is wrong if a paramedic even feels the need to seek this. Rather, the more appropriate behaviour is along the lines of a competent person who quietly, stoically gets on with the job and recognizes it as 'a privilege'. To be a good paramedic is to be an up-to-date clinician. It is to work carefully and compassionately with your patients and co-workers and to respect their autonomy and privacy. It is to get on with your duties without seeking approval. Don't boast. Don't wear a stethoscope around your neck or add some kind of Batman-style 'utility belt' to your uniform with various pouches and gadgets. Don't treat other ambulance workers as 'bag carriers'. You are allowed to be cynical, but never in front of patients. You will occasionally engage in a good moan (but only if you have put sufficient years into the job, and not so often that it drags others down). Your first priority is to treat your patients with care and dignity. A close second is to be considerate and supportive to your crewmates. Your own needs—and many of the requirements imposed by employers and managers—are not to be valued and prioritized.

As we shall see below, these informal rules pertain to individual responsibility and an appreciation that there exists a profession of paramedicine that needs to be contributed to and upheld. The profession's status and needs are of greater importance than those of the individual, and the profession's priorities and logics do not always neatly intersect with those of the employer. All professions—medicine, law, nursing, teaching—possess explicit and implicit norms about what represents appropriate individual behaviour and conduct (Liaschenko and Peter, 2004; Monrouxe et al., 2011; Smith, 2002; ten Hoeve and Jansen, 2013), and conscientious and respected paramedics will develop their own 'personal ethics' (Dallyn and Marinetto, 2022) that corresponds to these expectations.

Energy needs to be spent on developing a personal ethics by learning and reproducing formal and informal codes of conduct. Occasionally it means standing up for one's self and one's profession, to deflect or push back against unreasonable behaviours encountered at street-level. Several of the paramedics commented on challenges created when other clinical professionals don't understand the paramedic role or act in condescending or belittling ways towards those in green. Paramedic Practitioner Katie mentioned this to me several times:

'Some other, established professions can be difficult, they can look down on us. Not all of them by any means, but some individuals. Some nurses can be so snotty. They get grumpy when you bring in patients to their A&E. I'm sorry, but what are we supposed to do?! [. . .] The NHS is not integrated across all services, it is very messy and it feels like you are on your own. The district nurses get grumpy when you phone them.

For example, I was at a palliative care home. The patient had an infection but did not want to be transported to hospital. I got into an argument with the care home staff and the patient's GP. It was really hard to get her the care she needed. She needed hospital but didn't want to travel. But I also couldn't safely leave her there without at least a better assessment of the patient. She was complaining of severe throat pain and could not swallow. Her notes were a mess and it wasn't clear what medication she was on. After about twenty minutes I called the GP again, and her reaction was 'I'm not here at your beck and call.' Some of the doctors are spoilt, they are like borderline autistic in how they deal with you!'

Most of the paramedics would react to inter-service incivility with a shrug, just letting this sort of thing slide. Others would expend considerable effort explaining what paramedics can do, advocating for themselves and their patients in the complex NHS health economy and broader 'blue light' environment. Both reactions could be considered ways in which paramedics 'adapt and survive' in the pre-hospital environment. As ever, there were joking elements to this, with midwives labelled as 'madwives' and firefighters as 'water fairies'. Paramedic social media accounts will signal their dislike for the term 'first responder' and to educate people not to use the hated term 'ambulance driver'.

Paramedics, like any other uniformed occupation, will also have to stand their ground in the face of aggressive, unreasonable and insulting behaviour from the public, and even the possibility of having to confront the risks of physical violence. Often this comes in the form of drunk and disorderly patients. These are usually fairly easy to manage, but there is always the risk that events might get out of hand:

> David: I've had this a few times. There was this druggie who called me 'gay boy' and all that sort of thing. I was a probationer, a technician and it was probably the first time I'd had this. Sometimes you do have to restrain a patient more against himself than anything if they are just totally out of it. I once had a guy trying to strangle himself with the seatbelt in the ambulance and obviously you have to intervene and physically prevent them from hurting themselves. He was just sat there wrapping it around his own neck, it was bizarre. Sometimes you get alcoholics who are on a night out and totally drunk, sometimes they square up to you and you have to sort of calm them down and not get involved. I had two guys, Polish guys in Metroville who were totally out of it, booze and drugs. One is so high, he is hallucinating, very hyperactive, but his mate is also high but in a different kind of place, more in himself, kind of hunched over. But the hyper one then touched my crewmate, a female paramedic, he touched her arse! It was so weird he was then trying to lick her! The PPED just reacted, and said 'right that's it, I'm not having this', and we both grabbed him and pushed him down onto the trolley and strapped him very tightly, pulled the straps round and he was going nowhere! But there is so much fear of doing something wrong. There's CCTV in the ambulance, it can look bad if you are using any kind of force. Maybe the fear is exaggerated to some degree.

Paramedics would largely 'rise above' these kinds of incident and later 'laugh them off'. Nastier incidents I came across included examples of 'road rage', verbal abuse, being spat at, being physically grabbed and punched, and struck or stabbed at with household items. Some recalled dangerous and frightening incidents involving physical threats or assaults from patients or bystanders where paramedics had no route of escape, or situations where they got into physical confrontations with patients suffering from 'acute behavioural disturbance', who cannot be reasoned with and who are 'out of control'. Many have been to calls where a patient had 'kicked off' for largely unexplained reasons in any setting, including the hospital destination. News reports about ambulance crews being assaulted are quite

common. According to the GMB union, there are on average eight attacks on ambulance crews somewhere in the nation every day.[7] 70% of
paramedics who took part in a College of Paramedics survey claimed to
have feared for their own safety, while 80% had experienced verbal abuse
and 49% physical abuse.[8] Violence against medical staff can be extremely
upsetting and stressful, potentially leaving lasting psychological damage.
They are sometimes followed by years of background stress when they escalate into legal cases that involve ambulance staff being called as witnesses
in court proceedings.

The variety and unpredictability of street work exposes ambulance crews
to this, often when alone or in pairs, in settings considerably less safe than
those for clinicians in a hospital or a psychiatric unit. Violence is a threat
that all paramedics are likely to have experienced at some point, but it was
described to me as something rare and—strangely, it seemed to me—not
especially noteworthy. Such incidents might make a good ambulance station story, but they should not be used to try to obtain hero biscuits. In
the ambulance world, the 'professional' response to combative or insulting
patients is not to make too much of it. Instead, the best action is to simply
move on and focus on the next call where a patient might be in real distress and is deserving of the attention of a focused, caring professional who
is able to respond and whose efforts will be appreciated. As I will argue in
this next section, a genuine focus on patient need is a central element of
street-level professionalism. This is such a powerful factor of their identity
that they will tend to 'shrug off' stressful incidents and even violent ones
to an extent that risks disregarding their own health.

When the Clock Stops: Professionalism as Street-Level Care

The expansion of paramedic scope of practice has empowered paramedics
and their patients. Broadening the range of care that paramedics can
provide to patients in the field has improved the prospects for clinical outcomes and potentially obviates the need for stays at hospital that
many patients want to avoid. This focus on the patient is a key element
of paramedics' 'ability to respond', an ability that has been progressively
enhanced over time. But this ability does not operate in a vacuum.

---

[7] GMB report 'In Harm's Way', https://www.gmb.org.uk/sites/default/files/IN-HARMS-WAY.pdf
Accessed 11 May 2020.
[8] 'College of Paramedics' 2021 survey finds that nearly three-quarters of paramedics have feared for
their own safety of felt threatened at work', *Paramedic Insight*, 7(3): 14.

We have seen how the paramedics' ability to serve the public can be enhanced or hampered by broader social actions or inactions, notably those of the NHS trusts managing paramedics and the patients calling them. Classic studies in service and technical work have explored the 'service triangle' of worker, manager, and customer (Korczynski, 2003;Lopez, 2010). Each side of the triangle can potentially enhance or destroy the prospects for 'good service' to happen. It is a similar story in the abulance environment, with some key differences. When a vehicle is dispatched to a patient call, the 'management' side of the triangle is not always present or knowledgeable about the progress of that call. For long parts of most emergency calls, management is not present or even in direct contact with responding paramedics, meaning that for some time the emergency 'service' is formed not of a triangular, but of a diadic, relationship. The much sought-after situation of a 'good' or 'working' job—what I theorize as 'professionalism in process'-can potentially be manifested mostly by the efforts of the responding crew in ways that management can't intervene in, disrupt, or ruin. At the level of individual patient episodes, it is possible for professionalism in process to occur without management being an actor in that process; something along the lines of 'you and me against the world' in the words of Simon from Chapter 4.

It was Critical Care Paramedic Rob who helped me see this most clearly. For all the complaining about performance management and target response times, he spoke eloquently about care practices when a solo or dual crew arrives at scene and the clock stops. As he explains: 'it is then that you are in control':

Rob: For me, there is this very important middle bit, where you do get to strike up a relationship, when you do have time with the patient. You get a call, you are aware of the time target but even that does not really come into it. You get there, then you can take over, when you are at the scene with the patient, it is then that you are in control, you are managing a patient episode, it's as it should be. The clock is stopped at that point. You have autonomy at that point.

Rob's judgement that this is 'as it should be' is a powerful statement of the ethos of paramedic professionalism. The patient is at the centre of the focus of professional interaction. Ambulance crews are usually very attentive and compassionate, especially so at scenes that are clinically very serious, distressing, and sad. They expend effort in caring not only for the patients but also for family members and bystanders. The duty of explaining the process or outcome of a call can be exceptionally trying, especially when a paramedic has to convey the awful news that a patient has died or that the family should immediately start preparing for the worst. Born of hard experience, crews would do whatever they could to provide as much dignity as can be expected in the often strange and awkward setting of an emergency scene. For example, they would do whatever possible to improve the visual presentation of a deceased patient in the cramped and functional space of an ambulance interior. In less serious cases, ambulance crews were very often exceptionally skilled in social interaction, calming patients down, learning about their life, family, job, or hobbies, sharing a joke with them, finding ways to make the interaction more natural and more relaxing for all present. All of this is seen to be very much a part of paramedics' informal professional duties towards patients and the public.

Often the patient is probably not really 'deserving' of an emergency response according to clinical need in a highly rationed service. But, once summoned to a call, it still makes sense for the paramedic to give that person undivided attention, even if deep down the responder knows that clinical need does not really warrant it. A paramedic might 'go through the motions' somewhat when the callout is clearly unnecessary. I saw several incidents where crews would subtly indicate to one another privately that the caller was 'being a baby' or 'a timewaster'. Occasionally a patient might pick up on this. But with the patient at the focus of the service interaction, it makes no sense for paramedics to belittle or lecture the patient, or show their frustration in any way. That would be pointless. Most patient interactions are around an hour or so in length, meaning that each one on its own does not call for massive reserves of endurance. So why risk worsening the call by doing something that could lead to open conflict, unpleasantness or complaints? With 'rubbish' calls, the best reaction is to get through them in the present and laugh them off later if necessary. And even amid the 'rubbish', there is often some degree of patient need. In the example below, David describes a patient with ME who, while acting unpleasantly in parts of this episode, still represented someone in need of care. Paramedics are human and compassion fatigue is an ever-present danger. David admits to 'putting on a show of professionalism' and where 'a different side of him took over'. Paramedics need to be very careful about this.

David: I had a Chronic Fatigue Syndrome patient a few weeks ago. He had ME and was a repeat caller. Now he was claiming he can't get out of bed and had been in bed for a couple of days, unable to walk. But his story didn't make sense, he mentioned going to the shops, seeing the GP, so he had been walking. We asked him to sit up on the bed, he did this, and shuffle over to the end of the bed to get up. He started to walk about then did his 'collapse'. It was the most pathetic one I'd seen thus far. It was ridiculous. The way he went all wobbly first. A different side of me took over. I'll still be professional, I'll put on a show of professionalism, but I'll not be polite anymore. It's here where your crewmate can be helpful. He kind of took over from me 'coz he saw I was pissed off. We got him to the hospital and we had the trolley raised to the height of the bed and asked him if he could shuffle over onto the bed. Of course he says he can't, so we had to slide him over. At that point he then starts to be really abusive, slandering me, really! I laugh about it with you, you know, as it's so ridiculous! He was saying things like 'I beat you beat your wife and give her black eyes!' He called me a 'snotty bastard!' It was just so pathetic it didn't really upset me, you just let it slide off. Once back on the vehicle we were just laughing at him, it was like some ridiculous comedy act of his, the whole thing, the wobble and collapse, the abuse. It is actually pretty funny, and it becomes the source of jokes for the rest of the shift.

How a paramedic understands the meaning of dealing 'professionally' with patients is partly dependent on their individual reading of patient need, attitude and conduct (Seim, 2020: 92–5). This reading is also influenced by a crewmate as the patient is constructed by the crew as more or less 'deserving'. It is here where the considerable autonomy of ambulance work can potentially be abused.

Another highly experienced paramedic with managerial and educational responsibilities spoke about the 'drama' and even 'comedy' involved and the need 'to play a role' when an ambulance vehicle shows up 'on scene'. His account mentioned the long-running BBC TV drama series *Casualty* which is filmed in one of the regions where he works. He noted that paramedics play various 'theatrical' roles, an important aspect of their behaviour that has the potential to be misused. In his view, ambulance professionals should take care to play their roles in the right way, 'acting' professionally whenever in contact with the public, but also having the ability to 'edit' the roles, almost as in a dramatic repertoire:

Duncan: Your status as a paramedic from the outside, I mean in terms of the public, is almost godlike. You arrive on scene and its 'enter stage left', it's theatrical. We have BBC people down here for shooting episodes of *Casualty* and I said this to them. They didn't really get what I meant. Well, they wouldn't. Paramedics are bit-part roles in programmes like that, it focuses on the hospital; doctors, nurses primarily. I don't watch it. Anyway, there is a drama, a comedy sometimes even. I mean you have to play a role on scene - you need to be able to change your demeanour, your behaviour to suit the patient, even changing your voice (he puts on a working class, regional accent: 'you alright, mate?'). This has huge potential for abuse. You can get a bit judgemental. We are close to police, they are as autonomous as we are.

During my observations I was often struck by how casual ambulance crews seemed about new calls coming in. The MDT would sound an alert, or a radio would bleep with a new call and I would jump somewhat, and sometimes feel a bit of dread. The crews, on the other hand, seemed to feel nothing. In fact, their first reaction was often to joke about it, and mentally downgrade the call, downplaying the sense of emergency. The feeling was 'let's just see what it is when we get there'. Assumptions are made about the details of the call. Often the symptoms don't seem to make sense, or indicate that the job is likely to turn out to be 'dross'. Another element to this was continual joking about the exaggerated drama that sometimes accompanies the act of calling an ambulance. A common reference to this would be to the 'over-excitable first aider' who has been involved in the call and is waiting for the ambulance to arrive. Often the 'first aider' will make exaggerated arm movements towards the ambulance when it arrives, as if calling down a rescue helicopter to a risky landing.

In a similar way to that reported by Seim (2020), ambulance crews also made assumptions about the call based on its location. Some areas are notorious for poor quality housing, poor general health, and high crime rates. I recall being directed to a standby post in an area where the paramedic commented that the calls tend to be to 'people with hoopy earrings and furry tracksuits'. The following is a fieldnote observation where one responder explained ambulance work to me in probably the most uncompromising way I came across, in a way not dissimilar from how other uniformed occupations such as police and fire can express harsh views in private about a public they see as underserving, disrespectful or just plain stupid (Chetkovich, 1997; Moskos, 2009; Waddington, 1999).

It's hard, you have to not make a judgement. You get the address, the age, and you're thinking 'smackhead'. In the back of my mind, I'm thinking 'do I really put my foot down for another dead heroin addict?'

[. . .]

Every region has its timewasters, every Trust has its frequent flyers, you know? We're often spending longer at scenes debating what to do with them. It can be a long discussion, and people can be very uncooperative, or just have no idea what they want. As the responder, you're also not sure. A lot of the time they refuse to go, but you encourage them to come because it's just that bit safer. There are often sensible reasons to leave at home, but with so many callouts it gets harder to know what is a real emergency. It's gonna be Captain Cockup one of these days. One of those timewasters are going to cost someone their license.

Then there's the benefits lot. They will just take, take, take, and they never contribute to anything. Their attitude is often just awful, you know what I mean? You have to be professional in front of them, but you're thinking, you're demanding this from me, and when have you worked? When have you paid any taxes? You think the world owes you a living, the real reason you're in the shit is that you've just done nothing to change it. Yeah, people do have shit to put up with, but do you know what I mean, you have to try to do something, you have to try to fix yourself up?

So much depends on context. A paramedic will attune their degree of patient care to where they think it is most needed. Emergency scenes can be busy and fraught. Calls to the aftermaths of fights or road accidents can be complex, because patients often view their injuries through a prism of blame. They can be indignant if they think a responder is paying comparatively more attention to one injured party than the other, especially if the person being treated appears to be in some way responsible for the fracas or traffic collision. But a paramedic is not interested in who might be responsible for the injuries, and will rightly triage treatment according to clinical urgency. This sometimes requires the 'management' of a scene, including of unruly or disruptive patients and bystanders. Paramedics tend to be very skilled at screening out distractions. In fraught or edgy scenes of this kind, bystanders on the fringe of the incident might approach the car or ambulance to make some kind of claim as to what happened or offer their view on what should happen next, but a paramedic will usually make it clear that this input is not welcome, perhaps by closing the door or winding up the window. Acting professionally involves an informal, personal calculus of where best to expend effort. Sometimes it means giving undivided attention to every patient even when there are probably other patients out there right now with higher need but waiting on a response. Acting professionally sometimes also means finding ways to avoid being dragged into things that are clinically irrelevant. It also entails 'decompressing' and 'blowing off steam' in private spaces, sometimes involving harsh judgements about patients and the public.

Paramedics can have quite high expectations for patients to present with the 'right' or 'appropriate' problems. They expect 'management' in the form of 111 and the EOC to filter out the least appropriate calls. And they also expect their co-workers on the street to behave in the correct manner. While ambulance crews do not look favourably on what they consider to be inappropriate calls, they generally act with kindness towards the patient. If patients get a sense that their conditions weren't taken seriously or that the responding crew lacked empathy, then official complaints might arise which, given the troublesome 'blame culture', are very much something to avoid. Along with long waits for crews to arrive at the scene of a call, complaints about unsympathetic or uncaring crews represent the main focus of patient complaints. Below, Rob explains how quickly a new recruit can become desensitized to patients' viewpoints of emergencies. Patients' own experiences of sickness, injury, and anxiety must be taken seriously, even if that is sometimes done with a somewhat superficial 'display' of kindness and empathy from the responding crew:

> Rob: I worked with a guy who was a new recruit, not long out of his degree. I met him a few months ago. He was so enthusiastic, so up for it, so ready to get out there and work and to learn more. Really upbeat, really clinically focused, really keen. But I worked with him last night and he'd already changed. Hands in pockets, huffy and blowy, grumpy. I said: 'Mate, there's no reason to be like that', we had a great shift, nothing at all taxing, nothing interesting, the most 'serious' was a broken ankle, and we dealt with that, the patient was great. I said: 'Look, I know this I absolutely nothing for you and me, but it is a very big deal for him, and we have to look at it that way.'

One way to make sense of street-level professionalism is to adapt Korczynski's (2003) and Lopez's (2010) notions of a service triangle to encompass an 'emergency service triangle'. The three sides represent the paramedic, the patient and the EOC. Callouts don't happen unless the patient's call is deemed important enough by EOC to dispatch a vehicle. But, having arrived at a scene, then paramedics' own behaviour—represented by his or her attitudes, skills, and emotion management—represents the most powerful and influential of the three 'sides' of this 'triangle' in determining how well or how badly the call progresses. Education, protocols, and training are central components of paramedic professionalism but, on a daily basis, the most important ingredient in bringing about professionalism in practice is the personal conduct and attitude of the responding paramedic. Below, Simon reflects on this vital, yet informal element of the enactment of professionalism:

> Simon: It's like, can you hug a patient? It took me three years to learn that you can. Can you hug a patient, am I thinking as a paramedic or as the man underneath? What made me do that? Was it my training or my personality, my upbringing, the teaching of my parents? I remember discussing this through with a very experienced paramedic who works for Highdale Mountain Rescue. An example from a few years ago, there was a girl who had thrown herself out of a taxicab at speed. She was covered in grazes and bruises but the main thing was that she was distraught. She had thought that the taxi driver was going to harm her or rape her. She was extremely frightened, basically. We got there, I got her up into the ambulance, and she was sat on the trolley, I was opposite. She was in floods of tears, partly panicky, but mostly just very upset and vulnerable. What she didn't need was me doing checks, BP, O2 saturation, etc. what she needed was a hug. She was vulnerable, and she was in my space, I was in charge of the situation, the power differential is all in my favour. It's a thin line. But she was also looking to me to help her. I just leant over and gave her a big hug and I could feel some of the fear released immediately. As I did this I could see my partner looking behind over his shoulder, giving me that quizzical look 'everything ok?' and I just nodded, 'yeah, it's all ok'. I just gave her the care she needed. It's just *care*. We gave her the best care, we took her to A&E and the handover was great, the doctor was also very caring and helpful. It was lovely, it was a beautiful thing to be able to do. [. . .] It's just you as a person, the things you learn, your first girlfriend, how your parents have raised you, or just learning to do the right thing.

Simon and Rob's short phrases: 'It's just *care*', and 'It's as it should be' are interesting and instructive comments. They remind us that a genuine focus on the wellbeing of the patient is the central element of street-level professionalism. This everyday street-level professionalism can be subtle, intangible, and not immediately obvious. It involves having a sensitive and respectful orientation towards patients and towards support workers who are less qualified. It can be seen in micro-behaviours such as Simon's description in Chapter 3 of the handling of the vehicle's ignition keys as a way of indicating which of the crew should be attending or driving in particular cases. It can involve paramedics kneeling down when talking to a patient who is sat on a chair, getting to eye level, holding a patient's hand and taking some of the fear out of the situation, managing one's own emotions in order to help manage the emotions of others. Ambulance crews are also experts at manoeuvring people around in carry-chairs, trolleys, and stretchers, trying their best to ensure a patient is treated with as much comfort and dignity as possible. Not all of this emotional and manual work is valued clinically, but all of it represents a contribution to the enaction of a particular kind of expert, professional work held in high regard by the public and in very high demand.

When I first met ECSW Pete, he described the duties and responsibilities of working on an ambulance in an interesting and original way. In his account, professionalism perhaps transcends the boundaries of official role identity. Anyone on the ambulance has to be businesslike. And as 'a business', both the public and the government expect results out of the ambulance service:

> We are much more like a business now. We *are* a business. It's no good turning up just to look good. There's been more investment put in, they expect results. . .. Although, the cutbacks are coming too, the pressure is ramping up, there is no doubt of that.

Pete's interpretations were often robust and uncompromising. He was not a paramedic himself but, often working alongside CCPs, he often talked about the life-and-death demands of this environment. The accounts of the various people I met, observed and interviewed differed in interesting ways. Some emphasized clinical ability and qualification in their discussions of the meaning of professionalism. Others had a more earthy interpretation, focused more on basic human needs, compassion, and care. Others portrayed 'professionalism' as a counterforce to the procedural and uncaring 'management' that dominates the NHS. But for all their differences in emphasis and flavour, all of the accounts portrayed 'professionalism' as relating in some way to *providing what is needed given immediate patient and situational need*. Street-level professionalism is about having the right training, attitude, and competence to adapt to conditions and to be effective in the field. Finding a definition of the paramedic's USP that everyone agrees on is probably impossible, but Sarah was definitely onto something with her discussion of *that ability to respond*. For all the problems and frustrations that paramedics face, the 'good news' is that the ability to respond is likely to remain in very high demand. Indeed, paramedics' ability to respond is being drawn on in increasingly complex and varied ways, given the intensity and breadth of contemporary social suffering.

Conclusion

This chapter has explored a range of factors that both constitute and restrict the enactment of professionalism in the complex street-level world of ambulance response. Paramedics, like nurses before them (White, 2014), have experienced considerable success with their professionalization project, and the clinical scope and occupational identity of paramedics have been transformed since the days of 'ambulancemen' and the 1989–90 dispute. But the practical benefits of professionalism are not always clear. Paramedics now possess (to varying degrees) all of the trappings or 'traits' of a profession (Goode, 1957: 194; Leicht and Fennell, 2001: 26). But these traits are 'essentially arbitrary' (Evans, 2020: 360) revealing very little about the true nature of an occupation's work, conduct, and experiences. Rather than debating whether or not paramedics have—or will ever—become 'fully professionalized' (Etzioni, 1969: vii), it is more instructive to consider how often—or how rarely—paramedics get to contribute to the realization of professionalism in process.

With health systems in chronic crisis and care workers at breaking point, I argue that 'professionalism' is most meaningfully understood less in terms of the accumulation of traits, qualifications or managerialist performance indicators, and more in terms of a social process or condition that actors attempt to bring into existence through various behaviours, dispositions, and actions. If pre-hospital emergency care is to function effectively, especially when under such duress, then responding paramedics, ambulance trust employers, and patients themselves all need to be contributing in ways that allow professionalism in practice to emerge. Only then will a care episode usually fall into the informal category of 'a working job', and ideally result in 'a good outcome'. Professionalism in process can contribute to enhancing the dignity of the patient, the care worker, and the profession that he or she represents.

The epithet 'street-level' thus has a dual function in my characterization of paramedics as 'street-level professionals'. Not only does it illuminate the importance of discretion, autonomy, and improvisation as workers attempt to figure out how best to serve the needs of the public (Evans, 2010, 2016; Lipsky, 2010/1980; Maynard-Moody and Musheno, 2003), it also hints at the limitations of 'the street' (Cnossen et al., 2021) as a setting for the expression of more traditional notions of professionalism. Out-of-hospital clinical activity based on emergency callouts often lacks the scope and capacity to conform to classical ideals of professional work. Paramedics, lacking the ability to control or influence the type of work coming their way, strive with great difficulty to bring professionalism in process into being. Part of the problem is the extremely broad spectrum of call type that they are sent to. The professionalization project has tended to emphasize the upper end of the informal value hierarchy in its focus on the effective treatment of high-acuity patients, but there is no real prospect for the spectrum of paramedic work to shrink back to its traditional focus on life-threatening trauma, strokes, and CPR. Paramedics are increasingly confronting the changed reality that unplanned primary care is not an unwanted appendage to their work, but a core area of it. Although this is not welcomed by some, it is doubtless an area where paramedics can contribute hugely to the health economy as well as being a place where paramedics' professional competence can be expanded.

The paramedics' 'USP' seems unlikely to be fully clarified and resolved. It will always be subject to revision and adaptation. The renowned sociologist Amitai Etzioni noted in 1969 the 'source of tensions' facing occupations on their road to professionalism. Over forty years ago Donald Metz's groundbreaking ethnography of ambulance services argued that

> EMTs and paramedics are struggling mightily to overcome unnecessarily difficult conditions in carrying out their work. As structures slowly change and as ambulance work becomes more institutionalized, perhaps it will be easier for these people to exercise their talents.
>
> (Metz 1981: 192).

Even longer ago an English non-fiction writer, Jean Bowden, wrote of the ambulance field as a 'half-world of responsibility without recognition' (Bowden, 1963: 190), and described ambulance crews as 'so self-effacing that hardly one member of the public in a hundred could tell you what an ambulance man looked like, beyond the fact that he wears a uniform with a flat cap and arrives in an ambulance' (Bowden, 1963: 191).

Gender norms, skills profiles, and uniform codes change over time. Decades on from these early writings, ambulance workers are now more able to 'exercise their talents', particularly in a technical and clinical sense. But they still face extremely trying conditions, continue to struggle for recognition, and still possess that self-effacing, down to earth nature. As such, paramedics seem likely to remain 'always a profession in the making' (Nicklich et al., 2020). The final chapter of the book will offer some conclusions on this debate, suggesting that the paramedics' experience of substantial technical and expert improvement yet continual frustration and restriction is far from unusual. Paramedics are not alone. Rather, given the rising tides of neoliberal competition and managerialism, the paradox of being increasingly upskilled and in high demand, while simultaneously being overworked, mistrusted and maltreated to the point of personal crisis is actually a common feature of the professional experience of many occupations.

7

Professionalism as a Quest for Dignity

> 'We may not always get to perform the version of the job we
> envisaged, but we are participating in the communal matrix, and
> this is somehow the stuff of life.'
>
> **(Jones, 2020: 320)**

Why does professionalism matter? Why is it so sought-after as a prin-
ciple around which occupations aim to orient their work, duties and
identities? The most obvious answer is that professionalism provides an
occupation with autonomy, recognition, status and protection. Professionalism provides dignity. Working life is tough and working people will seek
protections and compensations that offset its tribulations.

This book opened with the 'Coach and Horses' industrial relations con-
frontation of 1989–90. Notions of 'professionalism' weren't often part of
the language of that conflict, but the dispute was most certainly about dignity. Ambulance workers sought recognition in the form of better pay and
conditions at a time when these had clearly fallen behind those of comparable occupations. The dispute, however, wasn't just about workers' pay and
conditions. The campaign demonstrated society's admiration and sympathy for 'ambulancemen', and it marked an important step on the long,
ongoing road to modernizing and clinically expanding what an ambulance
service can do for patients. It wasn't only the workers who stood to benefit
from an uplift of ambulance worker status. Patients could benefit, too.

The affordance of 'professional' status isn't something that an occupation strives to obtain only for self-interested reasons. Occupations believe
they possess an important service to offer society, and professionalism is
a way to find recognition for that service, and to put in place protections
so that only the right people with the proper qualifications, skills, bearing, and attributes are able to provide it. In this way, professionalization
can provide dignity not only for an occupation, but also for that occupation's clients. Professions are often thought of as powerful elements of 'civil
society'. At their best, they both reflect and contribute to society's broad

The Paramedic at Work. Leo McCann, Oxford University Press.
© Leo McCann (2022). DOI: 10.1093/oso/9780198816362.003.0007

aims to become better educated, safer, more compassionate, more technically competent, more rational, and more just. This short final chapter provides a conclusion to the book, in which I argue that the professionalization drive has had some important beneficial effects in further developing and cementing the value of the paramedic as a trusted and respected role in society. As we have seen throughout the book, the paramedics are a 'new profession' (Evetts, 2011), meaning they enjoy many of the classical traits of a profession, but not the status, protection, or rewards enjoyed by the most established, elite occupational groups. With life on the road being relentlessly tough and uncompromising, one might wonder quite what the professionalization drive has actually achieved. The chapter explores this question of achievement, arguing that professionalization hasn't managed to overcome many of ambulance work's challenges and limitations. Nevertheless professionalization has helped to cement the paramedics' status as very well trusted and respected. Professionalization confers several important social and cultural markers of status, and it sets the groundwork and direction for the further expansion of clinical knowledge that paramedics can apply in the field. The chapter moves on to speculate somewhat about the future of the field of pre-hospital care and of the paramedics working within it. It argues that ambulance roles, even if they still face uphill battles to achieve dignity and formal recognition, will always be invested with considerable levels of informal value, and that the professionalization project has played important roles in sustaining and upgrading that value. Neoliberalism and managerialism create incentives for many occupations to become 'new professions' while at the same time threatening potentially all professions with degradation. In a world where working life can be increasingly meaningless and precarious, the paramedic role remains a rare one worthy of the utmost respect and admiration, and the various features and meanings of 'professionalism' remain important resources through which paramedics can seek to defend and expand their dignity and the dignity of their patients.

The Changing Meanings of Professionalism

Academic, media, and policy discussions around the social meaning of professionalism have often devoted particular attention on the medical profession (DeAngelis, 2014; Freidson, 1970). I noted above a kind of equation of professionalism—the securing of professional status intends to confer benefit on both the occupation and the public. As regards medicine, the first part of that equation—that professionalism provides dignity to clinicians—has gradually waned in importance. Obviously doctors have long enjoyed an elevated status in society. But critics have increasingly noted how patient dignity is not always best served by the ways in which the profession of medicine operates, suggesting that members of the medical profession are overpowered in relation to their patients. In keeping with developments across other occupations (Leicht and Fennell, 2001), that power balance is starting to shift. The doctrine of patient-centred care (Berwick, 2009; Gaines et al., 2014) is an important marker of this.

Patient advocates argue that doctors should orient themselves towards their patients 'not as hosts in a care system but as guests in their lives' (Berwick, 2009: w559). Similar ideas are in play when ambulance crews talk of the 'privilege' of serving the public in times of acute need. Patient- or person-centred care implies stripping some of the traditional powers from the clinician and handing them to the patient (McCormack et al., 2021). No longer should doctors, nurses, or hospitals dictate to patients what is in their best interests. That would represent a very traditional notion of professional status; technocratic, paternalistic, undemocratic, and belittling. At its worst that form of care delivery is not compassionate. Patient-centred care, on the other hand, would be highly individualized, would involve patients and family members designing their own care plans and participating in rounds, and would see an end to restrictions on hospital visits (Berwick, 2009). The patient would be a consumer of healthcare services, encouraged to articulate their own needs, and given the choice to select or reject the care packages offered by various healthcare providers in a competitive, customer-responsive, industry.

Professionals will be resistant and sceptical towards much of this doctrine. They will argue that patients will ask for unnecessary or inappropriate procedures and treatments, and will refuse to comply with others that might be clinically correct. They point out that many patients don't want to be empowered, will reject the identity of 'consumer', and will expect an expert clinician to make diagnoses and decisions for them. They will suggest that a patient-centred relationship is dangerously close to an unfettered consumerist relationship in which clinicians with huge experience and a string of advanced degrees have no greater claims to expertise than any other person (Nichols, 2017). All of these objections are important.

But to frame this debate solely in terms of 'professionalism' versus 'consumerism' would be unhelpful. Patient-centredness isn't as far from the classical spirit of professionalism as it might first appear. Professions have always talked of a duty to their clients (Freidson, 2001; Leicht and Fennell, 2001; Muzio et al., 2019). Indeed, the promise that a particular occupation's skills, knowledge, and abilities are best placed to serve the public are central to the 'offer' they make to society when seeking a licence to practise. Professionals sometimes might not approve of the behaviour, requests, and conduct of their clients, and they are sometimes right not to agree with all of their demands. But professionals—like street-level bureaucrats—usually do expend great efforts in trying to achieve the interests of their clients. In some ways, the act of declining some of their pointless, unrealistic, or self-harming demands can help them to focus on more important needs that clients can't or won't recognize (Maynard-Moody and Musheno, 2003).

Berwick's (2009) paper sketches out what he self-consciously describes as an 'extremist' position on patient-centredness. Some of it is quite unrealistic and its championing of consumerism is often problematic. He says nothing about patients' radically different access to power and influence, especially patients' huge differentials in financial, social, and cultural capital. The NHS is officially free at the point of use but its resources are chronically overstretched and widely overused and misused. 'Patient choice' is difficult to envisage in a single-payer system, especially amid the constant rationing of scarce resources when providers are buried in an avalanche of patient demand. NHS services involve endless negotiations of need; some patient needs are reconciled quite easily, whereas others can be extremely difficult to address, or are unrealistic or contradictory.

Patient-centredness is largely a worthy approach, and its influence is here to stay. Reforms of the NHS have often talked of putting the patient 'at the centre' of healthcare. The development of market mechanisms in the NHS, and the focus on response time targets and maximum hourly and weekly waits for treatment clearly speak to these commercializing norms (Hyde et al., 2016). Professionals employed by large organizations are overseen by managers who are very sensitive to customer needs and the imperatives of audits and inspections. In the healthcare world, this tends to mean that patients' or managers' needs are placed higher on the list of organizational priorities than those of front-line employees. This can all be seen as part of the gradual shift in power relations between professionals and consumers. Consumers don't get everything their way, but their power relative to professionals is probably climbing (Leicht and Fennell, 2001). How responsive a healthcare profession is towards its patients is, therefore, an increasingly relevant and influential marker of professional status to place alongside the more traditional indicators at play in the clinical occupational hierarchy, such as its scope of practice, its depth of formal education, and the esteem in which its research journals are held.

Professionalization as an Unending Journey

Paramedics are typically highly regarded by their patients. Operating on a response basis and usually facing only one patient at a time, paramedics are very well-placed to contribute to the concepts and practices of person-centred healthcare (McCormack et al., 2021). But, as a relatively small and new professional group, they continue to be somewhat neglected in the wider NHS and are often excluded from policy discussions. Perhaps paramedics' sensitivity to patient-centredness could be leveraged by their representative professional body, the College of Paramedics, as part of the further expansion of paramedic practice and professional status? In expanding what they can do for patients in unplanned primary care, they could continue to outgrow their historically constrained role as 'physician-extenders'. The professionalization agenda can play an important part in expanding what paramedics can do for patients in the field which should simultaneously boost the social standing of paramedics themselves.

We are clearly seeing their scope of practice continuing to expand, both in the UK and in other nations. We can expect this to continue, particularly as paramedics are a relatively inexpensive clinical resource. New technologies and drugs will emerge. More protocols will be developed for paramedics to use. But the setting of ambulance work remains a complex issue for the paramedic field to negotiate. The expansion of ambulance provision into unplanned primary care runs counter to the traditional USP of paramedics as an emergency service with 'that ability to respond', and risks degrading some areas of paramedics' intangible and rare abilities to work effectively at street-level. And that context of 'the street' is also problematic. The street is not an easy place for professional work to unfold, and its logic of operations and response do not readily speak to the traditional markers of occupational respect in the NHS hierarchy which tend to circulate around the ability to practice advanced clinical procedures. Paramedics have made great strides in recent decades, but it seems sensible to assume that the 'new profession' (Evetts, 2011) of paramedicine will remain 'always a profession in the making' (Nicklich, 2020).

That there will be further changes is not in doubt. All caring and clinical professions face changing relationships to their clients due to rising and changing demand, the expansion of marketization and 'patient choice', the increased expectations that come with patient-centred approaches, and the rise of managerialist structures, practices, and audits. Other emergency services organizations are also undergoing change. The role and identity of the police officer has become very uncertain as part of the long-term process of police forces moving from 'fighting crime' to 'evidence-based practice' and 'dealing with vulnerability' (Bacon, 2019; Charman, 2017; Tomkins and Bristow, 2021). Fire and Rescue organizations have also been through changes, largely due to falling numbers of fire calls, as witnessed in the rise of community fire stations, retained crews, and experiments with co-responding to non-fire related health emergencies (McCann and Granter, 2019). Local and central government budget cuts have led to experiments with 'blue-light integration', including police, fire, and ambulance staff working out of shared premises and developing shared operational doctrine (Wankhade et al., 2019). Emergency services face changed operating models, such as wider use of volunteer responders, working on a kind of 'gig economy', or platform model, 'booking on' to their shifts using smartphone apps similar to those used by Uber or Lyft.

Paramedics have a duty to act professionally. But the enactment of professionalism does not occur in organizational and social vacuums. Others have duties, too. If the paramedics are ever to realize a 'truer' form of professionalism, then the workplaces that employ paramedics need to provide an employment culture that offers professionally acceptable conditions. Patients and bystanders should also act responsibly when calling for help and when involved in episodes of treatment, assessment, and transportation. If professionalism entails serving the public good and the raising the standards of social life, then all actors involved in its enactment (paramedics, employers, patients) need to act with civility, respect, and decency. Sometimes this does indeed occur. But there is room for improvement across all three dimensions.

Large portions of this book have documented the severe personal challenges of paramedic work. But it is important to acknowledge that paramedics' struggles with workload, wellbeing, excessive managerialism, and a lack of recognition and understanding of their work do not constitute evidence of their having somehow failed in their mission to become a profession. Many established professions are struggling with precisely the same issues. Professions such as academia (McCann et al., 2020), clinical psychology (Hannigan et al., 2009), architecture (Bowen et al., 2014), 'knowledge work' (Styhre, 2017) and policing (Lennie et al., 2020), feature staff burning out under workload and time pressure, intrusive managerialism and commercialism, restrictions on autonomy and control, and a growing sense of confusion about what constitutes 'good work' and how it should be conducted, measured, and recognized. In an instructive turn of phrase, schoolteachers are demanding 'professionally acceptable' rather that 'extreme' workloads (Greer and Daly, 2020).

One upside for the paramedics is that at least it cannot be said that their profession is at risk of erosion due to a lack of demand or a bad reputation. Their unique skills, their very high availability and the widespread respect they enjoy among the public means that the profession is increasingly in demand, increasingly visible, and increasingly seen as an attractive option for young people seeking meaningful and secure employment. NHS trusts face continual vacancies and shortages, and paramedic science degrees are oversubscribed, to the point where a newly graduated paramedic student will have little trouble finding a job. The issue isn't finding a job in this career, it's more about surviving it for any length of time before the stresses and strains become overwhelming.

Paramedics are not losing status because they are increasingly irrelevant, or that no-one is calling on them. Quite the contrary. Their problems are mostly derived from unmanageable demand for their services. One might hope, therefore, that 'the market' for 'expert service providers' would 'correct', and that the employment conditions for paramedics would improve to the point where their working lives become more sustainable. But there seems to be few signs of this happening, especially in NHS ambulance trusts. Widespread burnout, chronic sickness absence, and decisions to leave for other employers or quit ambulance life for good are leading to major shortages and vacancies for paramedics in NHS ambulance services.[1] Chapter 5 of this book explained the importance of intrinsic and unavoidable stressors of emergency response work that are partly responsible for burning out its staff. But there are other reasons, too, some of which are more readily addressable.

---

[1] 'Ambulance crews risk burnout with 1,000 posts vacant', *The Guardian*, 1 April 2018; 'Rate of paramedics leaving ambulance service nearly doubles', *National Health Executive*, 9 December 2015.

The Long Struggle for a Sustainable and Rational Emergency System

At the risk of belabouring the point or pouring cold water on paramedics' professional achievements over the last thirty years, it would be remiss not to mention the historical continuity of the stubborn reality of ambulance worker dissatisfaction. The work holds a very high intrinsic value, but it is relentlessly hard and chronically under-supported. Speeches in Parliament during the 1989–90 dispute regularly mentioned this, as Labour MPs would appeal to the toughness and high moral value of ambulance workers while trying to highlight the immorality of their having to work for free while locked out pursuing their industrial relations campaign. The following quotes are instructive. The details are dated, but the overall arguments made by Nigel Spearing, MP about ambulance workers as 'dedicated professionals' remain current:

> No ambulance worker is on strike. The ambulance workers have been locked out throughout the country. They are not taking strike action and, where they are allowed to do so, they are working without pay to provide a service which, as dedicated professionals, they are committed to give. [...][2]

And many of those in today's green uniform would surely recognize 'maladministration' on a 'grand and terrible scale':

> The total number in the London ambulance service is 2,379. In the two years 1987 to 1989, no fewer than 572 people left the service. Some 166 people retired, but of that number 148 had to leave on medical grounds. Only 18 people in two years retired at the normal retiring age. The pressures on the crews—psychological, physical and financial—were and are too great. Coupled with that is maladministration of the service on a grand and terrible scale.[3]

Jeremy Corbyn had the following to say:

> Ambulance workers suffer such trauma and tension all the time, yet they are told by the Secretary of State that they are no better than unskilled taxi drivers.

[2] *Hansard*, HC Deb 20 December 1989, vol. 164, c511.
http://hansard.millbanksystems.com/commons/1989/dec/20/ambulance-dispute
[3] *Hansard*, HC Deb 20 December 1989, vol. 164, c517.
http://hansard.millbanksystems.com/commons/1989/dec/20/ambulance-dispute

These are the people who deal with care and concern with every patient—the drunk picked up off the pavement on a Saturday night and the child in a road accident are treated with equal compassion.[4]

His examples betray a degree of exaggeration, but the essence of his rhetoric is important. It speaks to what a society would like to think about its 'mercy' staff. It is not dissimilar to popular discourses surrounding the noble sacrifices made by ambulance 'heroes' or 'angels in green'.

Society values paramedics highly. But laypersons seem largely unaware of quite how undervalued paramedics can be made to feel by management and by patients. We know that supportive first line management plays a critical role in staff wellbeing. The good news, therefore, is that one of the biggest issues that demoralizes and burns out the paramedics can—at least in theory—be addressed. An ambulance service line manager can be compassionate, careful, and supportive in his or her behaviours, slowly helping to contribute to a change of culture. More can be done by employers and educators about the critical need for managers to act with compassion and dignity towards their staff. With concerted effort, the blame culture can be addressed and it can be replaced by a culture of clinical learning and debriefing. Call prioritizing has improved with ARP, but there is scope to improve it further. Policy changes could rein in the chronic mistrust and obsession with targets engendered by the ambulance audit culture.

For their part, patients can be, and often are, supportive and appreciative. Unlike managers, patients are not employees in an organizational hierarchy, so it is hard to see who bears any responsibility for monitoring their behaviour and encouraging them to be respectful towards medical professionals and emergency responders. Indeed, raised cultural expectations of customer service (Korczynski, 2007), a growing sense of anxiety and neediness born out of social atomization and individualism (Bauman, 2000; Bude, 2017), and the discourse of patient-centred care might all run the risk of encouraging patients to be unrealistically demanding, to call more often, and to be more likely to complain. Paramedics routinely protest that EOC and 111 assessments of the risks of patient calls are excessive, creating more unnecessary work and falsely raising the expectations and demands of the public, and there seems to be some scope for managers and commissioners of emergency services to influence patients by readjusting these systems.

---

[4] *Hansard*, HC Deb 20 December 1989, vol. 164, c521.
http://hansard.millbanksystems.com/commons/1989/dec/20/ambulance-dispute

But overall, the behaviour and expectations of the public are difficult for anyone to proactively 'manage' or even influence. Chapter 6 suggested that, for 'professionalism in process' to occur in the pre-hospital field, three factors need to work constructively together: the paramedic responder, the employer, and the patient. Paramedics are not going to experience 'working calls' on every shift, but there is always the hope that some of the shift will feature 'good paramedic work'. There is not much that can be done to manage patients' behaviour. But managers can and should be supportive. Individual paramedics can also contribute their part by working collaboratively with crewmembers and acting compassionately towards patients, while remaining phlegmatic and optimistic in their bearing.

Recently, UK ambulance service senior management, aware of the bad publicity around bullying, staff sickness, and organizational abuse, has increasingly focused on encouraging 'compassionate' leadership, and (at least rhetorically) is claiming to listen to staff concerns and suggestions from below. This apparent sea-change in tone of public statements by ambulance managers seems to some extent a reflection of a genuine attempt to improve poor working conditions and address long-standing issues of sickness absence, and a reaction to scandals about poor staff treatment and toxic cultures. It is also likely to be a result of a realization on the part of employers that the growing clinical scope of paramedics mean they now have realistic prospects to outgrow their restrictive employers and find less fraught working environments. In the UK, the earning of academic degrees and the broadening of clinical scope has given paramedics professional mobility. NHS trusts have little trouble recruiting newly qualified paramedics, but are struggling to retain them.

Problems in the service over unmanageable operational pressure, long delays, unsupportive management, and staff burnout remain chronic. The Covid-19 pandemic has further exposed the vulnerabilities of the underfunded NHS system. Somehow, the NHS finds ways to battle through these crises and shortages, but at a high cost to the wellbeing of patients and staff. With all the focus on the immediate challenges of 'muddling through' (often by throwing all available resources at short-term objectives to fend off a collapse in performance management indicators), little or nothing is done strategically to help the system adapt to longer-term changes in operational scenarios. Health economies are changing. The ambulance service has drifted well away from its home waters of emergency trauma response. NHS ambulance trusts are now expected to expertly handle all of their traditional workload as well as a rising tide of unplanned primary care and psychiatric calls. The body that represents their senior leadership, the AACE, will proudly mention the scale and scope of the work that the service handles, but is wary of raising concerns about whether this vast workload is sensible or manageable. Efforts to inform patients about the appropriateness of using the 999 service seem to have little effect. Patient demand outstrips supply not just for ambulances, but for A&E units, hospital wards, GP surgeries, community nursing, and psychiatric care.

There have been repeated warnings and inquiries about the ambulance service being overwhelmed (National Audit Office, 2017; House of Commons Public Accounts Select Committee, 2019) but, other than a redesign of the call prioritization systems under ARP, nothing substantial has been done to address this crisis. If society really is serious about having a sustainable, world-class emergency ambulance service, then it needs to consider the shape and design of it. One of the most central issues it needs to address is quite simply the question of 'what is an emergency?' If the ambulance service is so heavily overloaded then society needs to ask why this is happening. Weren't societies supposed to have become 'knowledge economies' of 'expert workers' (Bell, 1976; Kerr, 2001/1963)? Hasn't advanced technology and medical knowledge made society healthier? Why are so many people in need, and why is that need often so urgent?

Ambulance services are not the only public service providers that are badly overstretched. Overloading of one part of the welfare state leads to backlogs and overflows that others try to pick up. Chronic problems such as poverty, poor housing, stress, poor diet, social exclusion, substance abuse, and loneliness all contribute to the buildup of health-related problems. When primary care or community care is unavailable, then a rapid accumulation of 111 or 999 calls is inevitable and, over time, becomes increasingly normalized. Ambulance services cannot refuse these unplanned primary care and psychiatric type calls, and paramedics can respond well to them. But this isn't a sensible use of an emergency response system. Earlier interventions need to be available to alleviate social suffering before emergency responses become the only option left. Prevention is better than cure. Effective change in this regard would require massive investment into all of the public service professions. It would entail major improvements to public health, primary care, hospital capacity, mental health provision, and a revolutionary improvement in the capacity of social care. It would also need to be accompanied by substantial changes to an economy and labour market which increasingly features zero-hours contracts, low pay, low intrinsic value, and chronic insecurity.

In the short term, maybe we also need radical change in the ambulance service to address its current, impossible, workload? With confusion elsewhere in the emergency services world, maybe there needs to be significant rethink of the 999 service? Can the ambulance service really run a time-critical response model to handle medical emergencies as well as unplanned primary care and psychiatric care? I'm not a clinician, healthcare economist or logistics expert, so I don't have the answers. But if I was to speculate, maybe the ambulance workload needs splitting into three separate divisions or sub-services according to type of need? There could be a paramedic-led ambulance service handling traditional, time-critical emergencies based on 'that ability to respond'. Alongside this, a fully-staffed and operational out-of-hospital unplanned primary care service could be built out of the 111 system—providing community nurses, paramedics, and care workers with vehicles to respond to the most urgent cases and the other, more chronic, cases can become part of their non-emergency caseload, in conversation with GP surgeries. Psychiatric calls could be handled on a similar basis—community mental health crisis teams are appallingly underfunded and overstretched, and there is surely room to expand their availability and mobility while also monitoring and treating patients with psychiatric disorders for longer periods, and with a broader range of interventions, especially longer-term talking therapies. It seems unreasonable to expect the ambulance service to handle all three areas of medical emergency, unplanned primary care, and psychiatric care. A major expansion of those latter two areas would also likely take considerable pressure off police forces, who are also regularly drawn into social care and psychiatric type calls and regularly complain that this isn't a sensible deployment of scarce uniformed resources.

Paramedics and other clinical professions are flexible and versatile. While inter-service rivalries will always exist, I've been struck by how many are not solely dedicated to their particular silo or uniform. They got into this business to help the public. As professionals they have shown they can think creatively, adapt to changing circumstances, and improvise through complex grey areas. Perhaps now it's time for governments and managers to do the same at a broader level. They have roles to play in rethinking and rebuilding the broader structures and organizing principles in which street-level professionals are employed so that they and their patients can be better served. Such action is long overdue.

Final Thoughts

Modernity offers the hope that society can become more efficient, more healthy, more civilized, and more just. Professions are central to that promise. It is vital that professions do not lose their aspirant role as independent actors in an open society, with contributing to the public good as their central purpose. In the paramedics' professionalization project, the patient—rightly—has been a central part of the focus on the steady expansion of clinical discretion. Educational expansion and curriculum development have played an especially important role here. But universities are not what they once were. They are also undergoing change, with their core mission now confused and distorted by market imperatives (Holmwood, 2016; Martin, 2016). Market incentives encourage a customer service logic and the recruitment of very large numbers of students to raise revenues. League tables and rankings encourage grade inflation to hit 'employability' targets. Those involved in paramedic education need to be wary of the incentives that universities have to 'dumb-down' paramedic science courses.

For all of the serious problems faced by paramedics and ambulance crews, there will always remain something very attractive about the calling. Despite everything, people working on ambulances still regard their work as 'a privilege' and 'the best job in the world'. There is huge public interest and sympathy for the identity of the paramedic. Paramedic science students are typically very keen to graduate, register, and start to work out on the road serving the public. Like other caring occupations, the paramedic role speaks to widely held ideas about social worth and dignity. If people find themselves facing a crisis or emergency, society expects that the person will receive rapid, compassionate, and effective care. When this doesn't happen, unspoken norms about a decent society are violated. This is why the media so often reports stories of unacceptably long waits for ambulances, overcrowded A&E units with patients waiting many hours in hospital corridors on trolleys and in wheelchairs, and describes the collapse of social care as a national scandal. The paramedic—like the police officer, nurse, midwife, doctor, or firefighter—represents a classic image of a special person that society turns to, to ask for help, to make sense of crisis, to use their expert skills and unflappable demeanour to provide immediate assistance. Society invests hope, trust, and value in that image. Something is very wrong in society when the volume of emergency demand so often outstrips supply. The ambulance world will always remain a place where professionalism, care, compassion, and dignity can be, and often is, manifested. More needs to be done by government—much more—if the dignity of paramedics and patients is to be further expanded and defended.

Notes on the Research and Writing Process

This book started life in 2009. Before then I knew nothing about the lives and work of paramedics. Why did I come to spend over ten years working on a book about them?

My research expertise lies in two broad areas: management and sociology of work. Since the early 2000s I have worked on various research projects about managers in general, and in middle managers in particular. These studies tended to ask similar questions about white-collar managers: Who are they? What do they do all day? And how do they feel about what they do? As part of an academic team, I began a research project in 2009 about the work-ing lives of middle managers in the National Health Service (Hyde et al., 2016). We were given access to study four NHS organizations as case studies, specifically a hospital trust, a mental health trust, a primary care trust, and an ambulance trust. I worked most of all on the ambulance case study and it was there that my interest in paramedics developed. Many of the managers I interviewed and observed were former or current paramedics. Talking to them and observing them as they went about their business, I became increasingly fas-cinated by the curious and unique ambulance world that they inhabited, and by the views they had developed about that world.

I scoured the academic literature on paramedic work (finding little of note back then beyond the classic US references of Mannon, 1992, Metz 1981, and Palmer 1983). I devoured the colourful and often disturbing content of the blog-based books written by NHS paramedics (Farnworth, 2020; Gray, 2007; Reynolds, 2009, 2010; Walder, 2020). Sev-eral papers on ambulance services emerged from the NHS managers project (Granter et al., 2019; McCann et al., 2013). But I knew there was a deeper story that no-one had really told thus far, a story about the changing culture of the NHS ambulance service and the broader paramedic world as it moves from an occupation to a profession.

During discussions with paramedics I was struck by a powerful tension. Our character-ization of their working world was largely downbeat, and I perceived that the ambulance staff appreciated the honesty and realism. But they also regularly pointed to aspects of the pre-hospital environment that had improved in recent years, noting that the standard of care that a paramedic can today provide to a patient has never been so advanced. They spoke with pride about how their clinical profession had developed, suggesting that the clinical elements (if not the operational and managerial) had improved out of sight compared to 10 or 20 years ago. In short, the professional power and status of paramedics were improving quickly, but much of their struggles and frustrations around the ways they were employed and managed remained severe and, in some ways, worsening.

Their own complaints to senior managers constantly fell on deaf ears, and I got a sense that they were pleased to have found some independent academic voices who might validate and broadcast their concerns. On several occasions I was told that 'the higher ups' should hear what we have to say. One particular paramedic encouraged me to send some of our findings and interpretations to a senior figure in an ambulance trust who would be likely to be receptive.

This senior manager turned out to be an extremely impressive person who lived and breathed paramedicine. We spoke for about 45 minutes during a busy day in which a personal assistant had to almost drag him physically from our meeting and into some national-level conference call. The central focus of our discussion was the role of the paramedic in the wider pre-hospital environment. He strongly recognized all of the problematic employment, managerial, cultural, and operational issues that our ongoing research had claimed were restraining the growth of the nascent paramedic profession. But he also suggested there was room for more optimism than our writings had indicated. He suggested I could do further observational shifts, and he set in chain the procedure for clearing me for this.

Some months later, following a process of registration and background checks, I had completed the Trust's observer training. I signed forms guaranteeing that I respect the privacy and confidentiality of patients and do not mention any patient-identifying details in any research publications emerging from the periods of observation. In addition, my employing university required me to undergo a higher-level risk assessment and legal indemnity process in case my riding on ambulances and attending incidents involved me in any accidents, harm, or complaints.

NHS Trusts all have well-developed procedures for taking observers onto operational shifts. The practice of ambulance crews taking a 'supernumerary' person on a vehicle as an observer has long been part of ambulance trusts' training, development, and appraisal culture, and has now also become a central part of the education that paramedic science degree students receive. Procedures for taking another person along to observe an ambulance shift are also required for the (fairly regular) practice of taking TV crews and journalists onto an ambulance to make documentaries which seem to be on our screens pretty much daily, with varying degrees of editorial accuracy: *Ambulance; 999, What's Your Emergency; Helicopter Heroes; Code Red.* Other NHS staff interested in pre-hospital medicine such as nurses, doctors, and ambulance trust managers will also undertake observations. Local politicians have also been known to take part in observational shifts on occasion. The practice used to be called 'third-manning' and that phrase is still in circulation. Given the pressure for paramedic science students to complete the observation and in-practice parts of their degree education, these places are in short supply. For that reason, I was able to complete only six 12-hour shifts, which sadly meant I was unable to achieve the depth and breadth of immersion into a field that one would hope for in a traditional ethnography.

Although my time on the road with crews was not very extensive, I found the observation of shifts to be indispensable for my understanding and characterization of ambulance work and the paramedics' professional craft. Crucially, it enabled me to appreciate the complicated and unvarnished context that they operate in. I saw how paramedics interact in front of patients and bystanders—calm, professional, caring, down-to-earth—but also how they interact with one another in private settings—tired, agitated, complaining, and bantering. Ambulance ride-alongs had similar value to the 'go-along' method in sociology and anthropology (Carpiano, 2009; Kusenbach, 2003). Going out on the road enabled me to see the types of environment that ambulance services are frequently sent to work in. Often these areas are dingy and depressed: run-down shopping centres, dismal office complexes, unloved housing estates on the fringes of forgotten towns. Paramedics and techs would regularly comment about the sorry state of the neighbourhoods in which they are responding, and even about some of the hospitals where they would drop off their patients.

As this observation period ended, my connections with the paramedic world continued to broaden and deepen. I attended several College of Paramedic meetings, conferences, and CPD and training events, all of which afforded me further insights into this developing professional world. I went on to make extensive use of personal interviews with ambulance staff. Through onward references, and even through paramedics contacting me having seen me talk at CPD events, I built up over several years a group of interviews with 20 people, mostly road paramedics and a small number of senior managers. I interviewed the managers, such as Ben, Linda, and Michael, in their offices at various ambulance trust locations. I interviewed the operational paramedics off work premises, usually in coffee shops. A couple of the paramedics made amusing comments about shady meetings with unknown persons and the furtive disclosure of 'the reality' of paramedic life. The following was from my fieldnotes around my meeting with Simon:

> I emailed Simon a few weeks after the recommendation from Stuart to set up a time. Simon's replies and questions were amusing, particularly about the proposed meeting place and how will we recognise each other. He wrote something like: 'this is quite clandestine isn't it? I am thin and wear glasses and sometimes wear cycling gear. I will carry a copy of last Sunday's *Der Tag* newspaper.' I reply in kind: 'This is getting weirder, I'm also thin and wear glasses and sometimes ride a bike. I will bring some photocopies of the Pentagon Papers and we can swap briefcases.'

Wary of protecting privacy and encouraging them to speak feely, I did not make use of a voice recorder when interviewing the paramedics. Instead, emboldened by my experience of scribbling profusely in the back of ambulances or in hospital corridors, I instead made copious notes of our conversations, immediately typing them up into word-processed documents when I returned home. I appreciate the limitations of note-taking rather than using voice recorders and I acknowledge that the book relies on my reconstruction of description and spoken text from notes and from memory (Zussman, 2016: 441). On the other hand, I like the way in which note-taking assists with anonymizing and cloaking, and I wasn't comfortable with recording the interviews or employing a transcriber to listen to and transcribe this sometimes-sensitive material. Some of the paramedics I would meet and interview multiple times, especially Emma, David, and Rob. Table A.1 shows the list of interviews in anonymized form.

As the table shows, the persons I interviewed were mostly highly experienced, including three senior managers. Almost all of the paramedics I interviewed were aged 30 and over, meaning many had an IHCD training background, although several had at some point later completed a degree in paramedic science in addition.

All of the names are pseudonyms. I explained in full to all of them the purposes of the research, and they were happy to take part on condition of anonymity. I was aware of the ambulance service's long-standing issues around a culture of fear and blame. But I found all of the paramedics to be candid and open, comfortable in talking to me on the basis of an individual professional expressing their private, anonymous views. On just one occasion, a person emailed me later to ask me to remove any reference to comments about a certain matter as this person was worried it could identify them to their employing Trust.

For the ambulance crews that I accompanied on the road as an observer, they knew that the purpose of my observations was to write an academic book about paramedics and their work. Of course, while on the road I ran into dozens of other crews on station, at scenes or at hospitals. Occasionally they would ask who I was, and I would explain. When I provided the answer of an academic researcher, they almost always seemed intrigued and supportive. From the patient and bystander point of view, people seemed not to pay me much

Table A.1 List of interviewees in anonymized form

Name	Position	Years in service
Ben	Director of Operations	25
Carl	Locality Manager	21
Connor	Paramedic and Area Manager	12
David	Paramedic (Degree trained)	6
Duncan	Head of Education	20
Emma	Paramedic and Line Manager	20
Gareth	Paramedic (IHCD trained)	21
Helen	Training and education manager	7
Jamie	Paramedic	10
John	Senior Manager, Special Operations	20
Karen	Paramedic (IHCD trained)	12
Liam	Paramedic (Degree trained)	13
Linda	Senior Manager	5
Michael	Medical Director	8
Rachel	Paramedic (Degree trained)	6
Rob	Critical Care Paramedic	10
Sarah	Paramedic (IHCD trained)	25
Simon	Paramedic (Degree trained)	11
Steven	Consultant Paramedic	20
Stuart	Paramedic and Practice Educator	16

attention at all, or to assume I was a paramedic student. In a sense that was exactly what I was—I was there to observe and learn. The crews would give me a high-viz ambulance jacket with 'OBSERVER' printed on it which marked me out as in some way connected to the crews, although I clearly looked at least twenty years older than the typical student. The mobile, unstructured and very busy nature of ambulance work meant that it was not possible to pursue an informed consent process from each person whose presence, actions, speech, or appearance might possibly be portrayed in my study. This is a difficult, but not unusual feature of ethnographic fieldwork in many settings (Hammersley and Atkinson, 2007: 42–3; Loftus, 2009; 201-9). It meant that I had to be especially careful about what observations I made, recorded, stored, and wrote up. For this reason, patients appear in the book infrequently, and, where they do, I have used very hazy and generalized forms of description. Throughout the book, all names of people or locations are fictionalized and the events described occurred several years before this book was written and published. In order to further avoid the possibility of any person being identified, I have sometimes slightly obscured or merged some of the individuals who appear in the book, or have modified their comments about some of the incidents that I discuss, a technique known as 'masking' or 'cloaking' (Glozer et al., 2019).

All of the people I interviewed and all of the crews that I observed seemed happy with the inductive and qualitative approach I was taking. I was half-expecting them to be more attuned to positivistic research approaches. Much of the academic research in the ambulance field consists of Delphi studies or randomized control trials, mirroring the general style of investigation and knowledge set by the medical profession. On the other hand, other clinical occupations (notably nursing and midwifery), have always had an affinity for qualitative studies, and I was relieved to find that there was plenty of support and interest in the approach I was taking. When I explained to paramedics that I was a sociologist of work, a couple of them mentioned that they had taken sociology A-level. One even asked if my study was a 'Paul Willis, *Learning to Labour*, type thing'?

I came across the concept of the 'curse of the observer' a few times. It seems to be a long-standing notion in ambulance work, referring to a kind of 'Murphy's Law' in that when a shift had someone 'third-manning', it is a good bet that it would be a boring shift that makes the work look dull, easy, quiet, and maybe dumbed-down, the reverse of the intense colour, expertise, and drama portrayed in 'reality' TV such as *Ambulance*, or *999 What's Your Emergency*. This notion also appears in Corman's study of Canadian paramedics (2017b: 85, 106) and Seim's work in the USA (2020: 68). For an academic observer, the very existence of this 'curse' was interesting. Clearly the whole purpose of my being there was to observe and learn about the 'reality' of emergency response work. That requires receiving callouts, some of which might be 'genuine' emergencies. While a crew (and a researcher in this case) don't want to be 'run ragged' with a string of high-intensity, stressful calls, they also don't want to be sat around inactive, continually repositioned at different standby posts for no apparent reason, or sent to nothing but mundane calls with no clinical content that really did not warrant an ambulance response. For the observer to learn more, there is always the 'hope' that the crew being observed would be sent to some dramatic calls. Personally, this desire was restrained on my part. I was anxious about witnessing the most unpleasant calls where patients were critically ill and where I would not be able to do anything to help. The fact that the 'curse of the observer' exists as a phenomenon in the ambulance world speaks to a central paradox of the 'edgework' (Granter et al., 2019) of ambulance crews. This is the schizoid disconnect that all crews feel, a sense of being torn between the desire for 'real emergencies' that 'we are trained for' and 'where we can make a difference', and the desire for 'steady' shifts that are not stressful and overwhelmingly busy, and where nothing serious can go wrong that might have ramifications (a very distressing incident that could trigger a pattern of mental ill-health, an episode where a paramedic might have made an error leading to patient complaints or management inspections or suspensions, or a scene that is tragically, unforgettably sad).

Exploring workers and organizations under strain can also be a challenge for researchers themselves. Police, fire, and ambulance organizations are often not healthy workplaces, and researchers can be exposed to considerable ethical challenges in how to make sense of and present their research findings (Lee, 1995). The research process can also be potentially distressing, involving listening to or possibly witnessing painful emergency scenes, and acting as a sounding board for anxious, struggling employees (Warden, 2013). Occasionally I felt acutely self-conscious about being on scene in this rather artificial 'observation' capacity, being unable to assist, except for sometimes carrying bags or equipment to and from the vehicles when asked. I worried a great deal at first about 'getting in the way' of patient treatment, but this was never a problem. Most of the time I enjoyed the ride-alongs, and the crews always made me feel relaxed and welcome. But there were occasions—as experienced by others doing similar work (see O'Neill, 2001)—where I found the experience hard and wondered why I felt the need to do this kind of research. In near-total opposite fashion from the paramedics themselves, I'd be happiest when the calls that came in sounded non-serious and I would worry when travelling to a call where a patient is likely to be very sick. On arrival, things tended not to be as bad as I might have feared. Just watching the paramedic crews' skilful emotional management of themselves and their patients would help me relax. I imagine that student paramedics must go through similar anxieties, which fade as they get progressively acclimated to the experience. Even though I stayed out of everything as per the instructions and policies I had agreed to, I occasionally felt on some calls that I might be intruding and I would back away. I made a mental note to remember these anxious feelings and to let them play a role in reminding me of the importance of writing this book in as sensitive a way as possible.

Deciding what to eventually put in the book also required a great deal of reflection about its purpose, and about privacy, risks of harm, and ethical appropriateness. Ethnographic and autoethnographic literature reminds us that a researcher gaining access to human stories does not entail an automatic right to use them in any way they see fit (Hammersley and Atkinson, 2007; Tolich, 2010). I chose to omit some of the most emotionally difficult scenes I saw or heard about, as I felt there was little to learn by sharing them. Similarly, some of the harshest gallows humour was so stark that recounting it out of context would be gratuitous and might make the ambulance service seem uncaring and callous. And then there is the issue of portrayal and balance. Much of my data about paramedics' employment experiences (sickness, poor standards of management, a blame culture) is pessimistic and downbeat. I had to report these findings in depth. And yet I also felt it was equally important that the book was not dominated by horror stories of burnout and conflict. The work is exhausting, there are management problems, and ambulance staff are exposed to some terrible things. But there is an upside. Although all were tired and strained, nearly all of the paramedics I met over the course of the years writing this book also said that they loved what they do, and that the professional identity of the paramedic held great personal meaning for them.

As a researcher I had a duty to present all of the problems of work that the paramedics recounted to me. But I did not want to portray their world only in stark terms and I was wary of burying the good news about the growth of paramedic professionalism in the bad news of unsupportive managerial climates and chronic employee disaffection. The more I got to know the field, the more I wanted to support the paramedics in their quest for higher recognition and status, and not to focus only on their struggles. Over time I came to realize that the best way to do this was to try to capture as much of the spectrum of experience as I could, with the long, detailed, and reflexive interviews of the paramedics playing a particularly important role.

I regret that I couldn't do more observations. Ultimately, I hope that the combination of ethnographic observations with the extensive unstructured interviews provides a richness, vividness, and authenticity about the complex nature of this fascinating new profession. Whether I have achieved that or not is for the readers of this book to judge, especially for any paramedic readers. For those working out on the road—day after day, night after night—I really hope that I've got my portrayal about right.

References

Abbott, A. (1981) 'Status and Status Strain in the Professions', *American Journal of Sociology*, 86(4): 819–835.

Abbott, A. (1988) *The System of Professions: An Essay on the Division of Expert Labor*. Chicago, IL: University of Chicago Press.

Abbott, P. and Meerabeau, L. (1998) 'Professionals, Professionalism, and the Caring Professions', in Abbott, P. and Meerabeau, L. (eds), *The Sociology of the Caring Professions*, London: UCL Press, pp 1–19.

Ackroyd, S. (1996) 'Organization Contra Organizations: Professions and Organizational Change in the United Kingdom', *Organization Studies*, 17(4): 599–621.

Adams, T. L., Clegg, S., Eyal, G., Reed, M., and Saks, M. (2020) 'Connective Professionalism: Towards (Yet Another) Ideal Type', *Journal of Professions and Organization*, 7(2): 224–233.

Alcadipani, R., Cabral, S., Fernandes, A., and Lotta, G. (2020) 'Street-Level Bureaucrats under COVID-19: Police Officers' Responses in Constrained Settings', *Administrative Theory & Praxis*, 42(3): 394–403.

Allan, S. M., Faulconbridge, J. R., and Thomas, P. (2019) 'The Fearful and Anxious Professional: Partner Experiences of Working in the Financialized Professional Services Firm', *Work, Employment and Society*, 33(1): 112–130.

Alvesson, M. and Spicer, A. (2016) '(Un)conditional Surrender? Why Do Professionals Willingly Comply with Managerialism?', *Journal of Organizational Change Management*, 29(1): 29–45.

Anderson, E. (2001) *Code of the Street: Decency, Violence and the Moral Life of the Inner City*. New York: W. W. Norton.

Anderson, N. E., Stark, J., and Gott, M. (2020) 'When Resuscitation Doesn't Work: A Qualitative Study Examining Ambulance Personnel Preparation and Support for Termination of Resuscitation and Patient Death', *International Emergency Nursing*, 49: March 2020, 100827

Association of Ambulance Chief Executives (AACE) (2013) *UK Ambulance Services Clinical Practice Guidelines 2013*. Bridgwater: Class Professional Publishing.

Association of Ambulance Chief Executives (AACE) (2015) *NHS Ambulance Services: Leading the Way to Care*. London: AACE, available at: https://aace.org.uk/resources/leading-way-care/ (accessed 20/4/20).

Association of Ambulance Chief Executives (AACE) (2018) *Emergency Ambulance Response Driver's Handbook*. Bridgwater: Class Professional Publishing.

Association of Ambulance Chief Executives (AACE). (2021) Ambulance Services and the Pandemic – A Review of 2020-21. London: AACE, available at: https://aace.org.uk/resources/ambulance-services-and-the-pandemic-a-review-of-2020-21-aace-dec-2021/ (accessed 25/1/22).

Bacon, C. (2019) 'Beyond the Scope of Managerialism: Explaining the Organizational Invisibility of Police Work', in Wankhade, P., McCann, L., and Murphy, P. (eds), *Critical Perspectives on the Management and Organization of Emergency Services*. Abingdon: Routledge, pp. 107–121.

Barley, S. R. and Kunda, G. (2001) 'Bringing Work Back In', *Organization Science*, 12(1): 1–95.

Barley, S. R. and Orr, J. E. (1997) 'Introduction: The Neglected Workforce', in Barley, S. R., and Orr, J. E. (eds), *Between Craft and Science: Technical Work in U.S. Settings*. Ithaca, New York: Cornell University Press, pp. 1–19

Bauman, Z. (2000) *The Individualized Society*. Cambridge: Polity.

Bechky, B. A. and Okhuysen, G. A. (2011) 'Expecting the Unexpected? How SWAT Officers and Film Crews Handle Surprises', *Academy of Management Journal*, 54(2): 239–261.

Becker, H., Geer, B., Hughes, E. C., and Strauss, A. L. (1977) *Boys in White: Student Culture in Medical School*. New Brunswick, N.J.: Transaction Publishers.

Beer, D. (2016) *Metric Power*. London: Palgrave Macmillan.

Beer, D. (2019) *The Data Gaze*. London: Sage.

Bell, D. (1976) *The Coming of Post-Industrial Society: A Venture in Social Forecasting*. New York: Basic Books.

Bell, R. C. (2009) *The Ambulance: A History*. Jefferson, NC: McFarland & Company.

Berwick, D. M. (2009) 'What 'Patient-Centered' Should Mean: Confessions of an Extremist', *Health Affairs*, 28(4): w555–w565.

Bevan, G. and Hood, C. (2006) 'What's Measured Is What Matters: Targets and Gaming in the English Public Health System', *Public Administration*, 84(3): 517–538.

Bittner, E., (1967) 'The Police on Skid-Row: A Study of Peace Keeping', *American Sociological Review*, 32(5): 699–715.

Blaber, A. Y. and Harris, G. eds (2014) *Clinical Leadership for Paramedics*. Maidenhead: Open University Press.

Bleetman, T. (2012) *You Can't Park There! The Highs and Lows of an Air Ambulance Doctor*. London: Ebury Press.

Bolton, S. C. (2001) 'Changing Faces: Nurses as Emotional Jugglers.' *Sociology of Health and Illness*, 23(1): 85–100.

Bourdieu, P. (1977) *Outline of a Theory of Practice*. Cambridge: Cambridge University Press

Bourdieu, P. (1990) *Homo Academicus*. Stanford, CA: Stanford University Press.

Bowden, J. (1963) *Call an Ambulance! The Story of the London Ambulance Crews*. London: Robert Hale.

Bowen, P., Edwards, P., Lingard, H., and Cattell, K. (2014) 'Occupational Stress and Job Demand, Control and Support Factors among Construction Project Consultants', *International Journal of Project Management*, 32(7): 1273–1284.

Boxall, P. and Macky, K. (2014) 'High-Involvement Work Practices, Work Intensification and Employee Well-Being', *Work, Employment and Society*, 28(6): 963–984.

Boyle, M. V. (2002) '"Sailing Twixt Scylla and Charybdis": Negotiating Multiple Organisational Masculinities', *Women in Management Review*, 17(3/4): 131–141.

Boyle, M. V. (2005) '"You Wait Until You Get Home": Emotional Regions, Emotional Process Work and the Role of Onstage and Offstage Support', in Hartel, C., Ashkansay, N. M., and Zerbe, W. (eds), *Emotions in Organizational Behavior*, New York: Psychology Press, pp. 45–65.

Boyle, M. V. and Healy, J. (2003) 'Balancing Mysterium and Onus: Doing Spiritual Work within an Emotion-Laden Context', *Organization*, 10(2): 351–373.

Braedley, S. (2015) 'Pulling Men into the Care Economy: The Case of Canadian Firefighters', *Competition & Change*, 19(3): 264–278.

Brehm, J. and Gates, S. (1997) *Working, Shirking and Sabotage: Bureaucratic Response to a Democratic Public*. Anne Arbor, MI: The University of Michigan Press.

Brent, S. (2010) *Nee Naw: Real Life Dispatches from Ambulance Control*. London: Penguin.

Brewis, J. and Godfrey, R. (2019) 'From Extreme to Mundane? The Changing Face of Paramedicine in the UK Ambulance Service', in Wankhade, P., McCann, L., and Murphy, P. (eds), *Critical Perspectives on the Management and Organization of Emergency Services*. Abingdon: Routledge.

Brodkin, E. Z. (2008) 'Accountability in Street-Level Bureaucracies', *International Journal of Public Administration*, 31(3): 317–336.

Brodkin, E. Z. (2011) 'Policy Work: Street-Level Organizations under New Managerialism', *Journal of Public Administration Research and Theory*, 21(2): i253–i277.

Brown, H. and Edelmann, R. (2000) 'Project 2000: A Study of Expected and Experienced Stressors and Support Reported by Students and Qualified Nurses', *Journal of Advanced Nursing*, 31(4): 857–864.

Bude, H. (2017) *Society of Fear*. Cambridge: Polity.

Caless, B. (2011) *Policing at the Top: The roles, values and attitudes of chief police officers*. Bristol: Policy Press.

Campeau, A. (2009) 'Introduction to the "Space-Control Theory of Paramedic Scene Management"', *Emergency Medicine Journal*, 26: 213–216.

Campeau, A. (2011) 'The Space-Control Theory of Paramedic Scene-Management', *Symbolic Interaction*, 31(3): 285–302.

Campeau, A. C. (2016) 'Paramedical Risk Framing during Field Referral for Acute Stroke and S-T Elevation Myocardial Infarction Patients', *Emergency Medical Journal*, 33: 414–417.

Carpiano, R. M. (2009) 'Come Talk a Walk with Me: The "Go-Along" Interview as a Novel Method for Studying the Implications of Place for Health and Well-being', *Health & Place*, 15(1): 263–272.

Chaney, P. K. and Philipich, K. L. (2002) 'Shredded Reputation: The Cost of Audit Failure', *Journal of Accounting Research*, 40(4): 1221–1245.

Charman, S. (2013) 'Sharing a Laugh: The Role of Humour in Relationships between Police Officers and Ambulance Staff', *International Journal of Sociology and Social Policy*, 33(3/4): 152–166.

Charman, S. (2017) *Police Socialisation, Identity and Culture: Becoming Blue*. London: Palgrave Macmillan.

Chetkovich, C. (1997) *Real Heat: Gender and Race in the Urban Fire Service*. New Brunswick: Rutgers University Press.

Chwastiak, M., (2001) 'Taming the Untamable: Planning, Programming and Budgeting and the Normalization of War', *Accounting, Organizations and Society*, 26(6): 501–519.

Clement, B. (2010) 'Introduction', in Conaghan, J. (2010) *Coach and Horses: My Story in the Ambulance Dispute 1989/90*, Merthyr Tydfil: Joseph Conaghan, pp. 9–20.

Clompus, S. R. and Albarran, J. W. (2016) 'Exploring the Nature of Resilience in Paramedic Practice: A Psycho-social Study', *International Emergency Nursing*, 28: 1–7.

Cnossen, B., de Vaujany, F.-X. and Haefliger, S. (2021) 'The Street and Organization Studies', *Organization Studies*, 42(8): 1337–1349.

Collins, D., Dewing, I., and Russell, P. (2009) 'The Actuary as Fallen Hero: On the Reform of a Profession', *Work, Employment and Society*, 23(2): 249–266.

Collinson, D. (2011) 'Critical Leadership Studies', in Bryman, A., Collinson, D., Grint, K., Jackson, B., and Uhl-Bien, M. (eds), *The Sage Handbook of Leadership*. London: Sage, pp. 181–194.

Conaghan, J. (2010) *Coach and Horses: My Story in the Ambulance Dispute 1989/90*, Merthyr Tydfil: Joseph Conaghan.

Corman, M. K. (2017a) 'Driving to work: The Front Seat Work of Paramedics To and From the Scene'. *Symbolic Interaction*, 41(3): 291–310.

Corman, M. K. (2017b) *Paramedics On and Off the Streets: Emergency Medical Services in the Age of Technological Governance*. University of Toronto Press.

Corman, M. K. and Melon, K. (2014) 'What Counts? Managing Professionals on the Front Line of Emergency Services', in Griffith, A. I. and Smith, D. E. (eds), *Under New Public Management: Institutional Ethnographies of Changing Front-Line Work*. Toronto: University of Toronto Press, pp. 148–176.

Coser, L. A. (1974) *Greedy Institutions: Patterns of Undivided Commitment*. New York: Free Press.

Coulson, A. (2009) 'Targets and Terror: Government by Performance Indicators', *Local Government Studies*, 35(2): 271–281.

Courpasson, D., and Monties, V. (2017) '"I Am My Body": Physical Selves of Police Officers in a Changing Institution', *Journal of Management Studies*, 54(1): 32–57.

Cousineau, M. J. (2016) 'Accomplishing Profession through Self-Mockery', *Symbolic Interaction*, 39(2): 213–228.

Currie, G., Burgess, N., and Tuck, P. (2016) 'The (Un)desirability of Hybrid Managers as 'Controlled' Professionals: Comparative Cases of Tax and Healthcare Professionals', *Journal of Professions and Organization*, 3(2): 142–153.

Currie, G., Richmond, J., Faulconbridge, J., Gabbioneta, C., and Muzio, D. (2019) 'Professional Misconduct in Healthcare: Setting Out a Research Agenda for Work Sociology', *Work, Employment and Society*, 33(1): 149–161.

Dallyn, S. and Marinetto, M. (2022) 'From Resistance and Control to Normative Orders: *The Wire*'s Cedric Daniels as an Ethical Bureaucrat', *Human Relations*, 75(3): 560–582

de Bruijn, H. (2010) *Managing Professionals*. Abingdon: Routledge.

de Rond, M. (2017) *Doctors at War: Life and Death in a Field Hospital*. Ithaca: Cornell University Press.

de Rond, M. and Lok, J. (2016) 'Some Things Can Never be Unseen: The Role of Context in Psychological Injury in War', *Academy of Management Journal*, 59(6): 1965–1993.

DeAngelis, C. D. (2014) 'Introduction', in DeAngelis, C. D. ed., *Patient Care and Professionalism*. Oxford: Oxford University Press, pp. xxiii–xviii.

Department of Health (DH) (2005) *Taking Healthcare to the Patient: Transforming NHS Ambulance Services*. London: Department of Health.

Desmond, M. (2007) *On the Fireline: Living and Dying with Wildland Firefighters*. Chicago: University of Chicago Press.

Dingwall, R. (1977) 'Atrocity Stories and Professional Relationships', *Sociology of Work and Occupations*, 4(4): 371–396.

Dixon, D. (2009) '"I Can't Put A Smiley Face On": Working-Class Masculinity, Emotional Labour, and Service Work in the New Economy', *Gender, Work & Organization*, 16(3): 300–322.

Docherty, T. (2014) *Universities at War:* London: Sage.

du Gay, P. and Pederson, K. Z. (2020) 'Discretion and Bureaucracy', in Evans, T., and Hupe, P. (eds), *Discretion and the Quest for Controlled Freedom*. London: Palgrave Macmillan, pp. 221–236.

du Gay, P. (2000) *In Praise of Bureaucracy: Weber, Ethics, Office*. London: Sage.

Eaton, G. (2019) 'Paramedic. *noun.*' British Paramedic Journal, 4(2): 1–3.

Eaton, G., Mahtani, K., and Catterall, M. (2018) 'The Evolving Role of Paramedics – a NICE Problem to Have?' *Journal of Health Services Research & Policy*, 23(3): 193–195.

Epp, C. R., Maynard-Moody, S., and Haider-Markel, D. (2014) *Pulled Over: How Police Stops Define Race and Citizenship*. Chicago: Chicago University Press.

Etzioni, A. (1969) 'Preface', in Etzioni, A. (ed.), *The Semi-Professions and their Organization: Teachers, Nurses, Social Workers*. New York: Free Press, pp. v–xviii.

Evans, T. (2010) *Professional Discretion in Welfare Services: Beyond Street-Level Bureaucracy*. Abingdon: Routledge.

Evans, T. (2016) 'Street-Level Bureaucracy, Management and the Corrupted World of Service', *European Journal of Social Work*, 19(5): 602–615.

Evans, T. (2020) 'Discretion and Professional Work', in Evans, T. and Hupe, P. (eds), *Discretion and the Quest for Controlled Freedom*. London: Palgrave Macmillan, pp. 357–375.

Evans, T. and Harris, J. (2004) 'Street-Level Bureaucracy, Social Work and the (Exaggerated) Death of Discretion', *British Journal of Social Work*, 34(6): 871–895.

Evetts, J. (2011) 'A New Professionalism? Challenges and Opportunities', *Current Sociology*, 59(4): 406–422.

Evetts, J. (2012) 'Professionalism: Value and Ideology', *Sociopedia.isa*. DOI: 10.1177/205684601231

Exworthy, M. and Halford, S. (1999) 'Professionals and Managers in a Changing Public Sector: Conflict, Compromise and Collaboration?' in Exworthy, M. and Halford, S. (eds), *Professionals and the New Managerialism in the Public Sector*. Buckingham: Open University Press, pp. 1–17.

Eyal, G. (2013) 'For a Sociology of Expertise: The Social Origins of the Autism Epidemic', *American Journal of Sociology*, 118(4): 863–907.

Farnworth, D. (2020) *999: My Life on the Frontline of the Ambulance Service*. London: Simon & Shuster.

Faulconbridge, J. and Muzio, D. (2008) 'Organizational Professionalism in Global Law Firms', *Work, Employment and Society*, 22(1): 7–25.

Fellows, B. (2020) 'Paramedics 50 Years On: 1970–2020. *Paramedic Insight*, 5(4): 10–11.

Fellows, B. and Harris, G. (2019) 'History of the UK Paramedic Profession', in Wankhade, P., McCann, L., and Murphy, P. (eds), *Critical Perspectives on the Management and Organization of Emergency Services*. Abingdon: Routledge, pp. 30–51.

Fellows, S. and Fellows, B. (2012) *Paramedics: From Street to Emergency Department Case Book*. Maidenhead: Open University Press.

Filstad, C. (2010) 'Learning to be a Competent Paramedic: Emotional Management in Emotional Work', *International Journal of Work, Organisation and Emotion*, 3(4): 368–383.

Fitzgerald, T. (2008) 'The Continuing Politics of Mistrust: Performance Management and the Erosion of Professional Work', *Journal of Educational Administration and History*, 40(2): 113–128.

Frankfurt, S. and Frazier, P. (2016) 'A Review of Research on Moral Injury in Combat Veterans', *Military Psychology*, 28(5): 318–330.

Freidson, E. (1970) *Professional Dominance: The Social Structure of Medical Care*. Chicago: Aldine Publishing.

Freidson, E. (1986) *Professional Powers: A Study on the Institutionalization of Formal Knowledge*. Chicago: University of Chicago Press.

Freidson, E. (2001) *Professionalism: The Third Logic*. Cambridge: Polity.

Furness, S., Hanson, L., and Spier, J. (2021) 'Archetypal Meanings of Being a Paramedic: A Hermeneutic Review', *Australasian Journal of Emergency Care*, 24(2): 135–140.

Gabbert, L. (2020) 'Suffering in Medical Contexts: Laughter, Humor, and the Medical Carnivalesque', *Journal of American Folklore*, 133: 3–26.

Gaines, M. E., Grob, R., Schlesinger, M. J., and Davis, S. (2014) 'Medical Professionalism from the Pateint's Perspective', in DeAngelis (ed.), *Patient Care and Professionalism*. Oxford: Oxford University Press, pp. 1–18.

Gascoigne, C., Parry, E., and Buchanan, D. (2015) 'Extreme work, gendered work? How extreme jobs and the discourse of 'personal choice' perpetuate gender inequality', *Organization*, 22(4): 457-475.

Gherardi, S. (2009) 'Community of Practice or Practices of a Community?', in Armstrong, S. J. and Fukami, C. V. (eds), *The SAGE Handbook of Management Learning, Education and Practice*. London: SAGE, pp. 514–530.

Ginsberg, B. (2013) *The Fall of the Faculty: The Rise of the All-Administrative University and Why it Matters*. Oxford: Oxford University Press.

Givati, A., Markham, C., and Street, K. (2018) 'The Bargaining of Professionalism in Emergency Care Practice: NHS Paramedics and Higher Education', *Advances in Health Sciences Education*, 23(2): 353–369.

Gofen, A. (2014) 'Mind the Gap: Dimensions and Influence of Street-Level Divergence', *Journal of Public Administration Research and Theory*, 24(2): 373-393.

Glozer, S., Caruana, R., and Hibbert, S. A. (2019) 'The Never-Ending Story: Discursive Legitimation in Social Media Dialogue', *Organization Studies*, 40(5): 625–650.

GMB (2018) *In Harm's Way: Confronting Violence against NHS Ambulance Staff*. London: GMB. Available at: https://www.gmb.org.uk/sites/default/files/IN-HARMS-WAY.pdf (accessed 25 June 2021).

Gofen, A. (2014) 'Mind the Gap: Dimensions and Influence of Street-Level Divergence', *Journal of Public Administration Research and Theory*, 24(2): 473–493.

Goode, W. J. (1957) 'Community within a Community: The Professions', *American Sociological Review*, 22(2): 194–200.

Graen, G. B. and Graen, J. A. (eds) (2013) *Management of Team Leadership in Extreme Context: Defending our Homeland, Protecting our First Responders*. Charlotte, NC: Information Age Publishing.

Granter, E., Wankhede, P., McCann, L., Hassard, J., and Hyde, P. (2019) 'Multiple Dimensions of Work Intensity: Ambulance Work as Edgework', *Work, Employment & Society*, 33(2): 280–297.

Gray, S. (2007) *Life & Death on the Streets: A Paramedic's Diary*. Cheltenham: Monday Books.

Grayling, A. C. (2017) *War: An Enquiry*. New Haven, CT: Yale University Press.

Green, F. (2004) 'Work Intensification, Discretion and the Decline in Well-being in Work', *Eastern Economic Journal*, 30(4): 615–625.

Greer, J. and Daly, C. (2020) 'Professionally Acceptable Workload: Learning to Act Differently towards Effective Change', *Impact*, 9: 15–18.

Griffith, A. I. and Smith, D. E. (2014) 'Introduction', in Griffith, A. I. and Smith, D. E. (eds), *Under New Public Management: Institutional Ethnographies of Changing Front-Line Work*. Toronto: University of Toronto Press, pp. 3–21.

Grint, K. (2010) 'The Cuckoo Clock Syndrome: Addicted to Command, Allergic to Leadership', *European Management Journal*, 28(4): 306–313.

Grint, K. (2020) 'Leadership, Management and Command in the Time of the Coronavirus', *Leadership*, 16(3): 314–319

Hällgran, M., Rouleau, L., and de Rond, M. (2017) 'A Matter of Life and Death: How Extreme Context Research Matters for Management and Organization Studies', *Academy of Management Annals*, 12(1): 111–153.

Hällgran, M., Rouleau, L., and de Rond, M. (2018) 'A Matter of Life and Death: How Extreme Context Research Matters for Management and Organization Studies', *Academy of Management Annals*, 12(1): 111–153.

Hammersley, M. and Atkinson, P. (2007) *Ethnography: Principles in Practice*. Abingdon: Routledge.

Hannigan, B., Edwards, D., and Burnard, P. (2009) 'Stress and Stress Management in Clinical Psychology: Findings from a Systematic Review', *Journal of Mental Health*, 13(3): 235–245.

Harris, G. (2014) 'Paramedic Leadership and the NHS Clinical Leadership Career Framework', in Blaber, A. and Harris, G. (eds), *Clinical Leadership for Paramedics*: Maidenhead: Open University Press, pp. 20–38.

Henckes, N. and Nurok, M. (2015) '"The First Pulse You Take Is Your Own – But Don't Forget Your Colleagues". Emotion Teamwork in Pre-hospital Emergency Medical Services', *Sociology of Health and Illness*, 37(7): 1023–1038.

Herbert, S. (2006) 'Police Subculture Reconsidered', *Criminology*, 36(2): 343–370.

Hewlett, S. A. and Luce, C. B. (2006) 'Extreme Jobs: The Dangerous Allure of the 70-Hour Workweek', *Harvard Business Review*, 84(12): 49–59.

Hockey, N. (2009) ''Switch On': Sensory Work in the Infantry', *Work, Employment and Society*, 23(3): 477–493.

Holdaway, S. (2017) 'The Re-professionalization of the police in England and Wales', *Criminology and Criminal Justice*, 17(5): 588–604.

Holmwood, J. (2016) ''The turn of the screw'; marketization and higher education in England', *Prometheus*, 34(1): 63–72.

Hood, C. (1991) 'A Public Management for All Seasons?', *Public Administration*, 69(1): 3–19.

House of Commons Public Accounts Committee (2017) *Ambulance Services Study Inquiry*. London: UK Parliament. https://publications.parliament.uk/pa/cm201617/cmselect/cmpubacc/1035/103502.htm

Hughes, D. (1980) 'The Ambulance Journey as an Information Generating Process', *Sociology of Health and Illness*, 2(2): 115–132.

Hughes, E. S., Dobbins, T., and Murphy, S. (2020) '"Going Underground": A Tube Worker's Experience of the Struggles over the Frontier of Control', *Work, Employment and Society*, 33(1): 174–183.

Huo, M.-L., Boxall, P., and Cheung, G. W. (2022) 'Lean Production, Work Intensification and Employee Wellbeing: Can Line-manager Support Make a Difference?', *Economic and Industrial Democracy*, 43(1): 198–220

Hupe, P. and Hill, M. (2019) 'Positioning Street-Level Bureaucracy Research', in Hupe, P. (ed.), *Research Handbook on Street-Level Bureaucracy: The Ground Floor of Government in Context*. Aldershot: Edward Elgar, pp. 15–30

Hyde, P., Granter, E., Hassard, J., and McCann, L. (2016) *Deconstructing the Welfare State: Managing Healthcare in the Age of Reform*. Abingdon: Routledge.

James, N. (1992) 'Care = Organisation + Physical Labour + Emotional Labour', *Sociology of Health and Illness*, 14(4): 488–509.

Johnson, T. J. (1972) *Professions and Power*. London: Macmillan.

Jones, J. (2020) *Can You Hear Me? An NHS Paramedic's Encounters with Life and Death*. London: Quercus.

Joseph, N. and Alex, N. (1972) 'The Uniform: A Sociological Perspective'. *American Journal of Sociology*, 77(4): 719–730.

Kahl, S. J., King, B. G., and Liegel, G. (2016) 'Occupational Survival through Field-Level Task Integration: Systems Men, Production Planners and the Computer, 1940s–1990s. *Organization Science*, 27(5): 1065–1341.

Kaufman, J. (2019) 'Intensity, Moderation, and the Pressures of Expectation: Calculation and Coercion in the Street-Level Practice of Welfare Conditionality', *Social Policy & Administration*, 54(2): 205–218.

Kellerman, B. (2018) *Professionalizing Leadership*. Oxford: Oxford University Press.

Kerr, A., and Sachdev, A. (1992) 'Third among Equals: An Analysis of the 1989 Ambulance Dispute', *British Journal of Industrial Relations*, 30(1): 127–143.

Kerr, C. (2001/1963) *The Uses of the University*. Boston, MA: Harvard University Press.

Kessler, I., Heron, P., and Dopson, S. (2015) 'Professionalism and Expertise in Care Work: The Hoarding and Discarding of Tasks in Nursing', *Human Resource Management*, 54(5): 737–752.

Khurana, R. (2007) *From Higher Aims to Hired Hands: The Social Transformation of American Business Schools and the Unfulfilled Promise of Management as a Profession*. Princeton, NJ: Princeton University Press.

Kilner, T. (2004) 'Educating the Ambulance Technician, Paramedic and Clinical Supervisor: Using Factor Analysis to Inform the Curriculum', *Emergency Medicine Journal*, 21: 379–385.

Klein, K. J., Ziegert, J. C., Knight, A. P., and Xiao, Y. (2006) 'Dynamic Delegation: Shared, Hierarchical and Deindividualized Leadership in Extreme Action Teams', *Administrative Science Quarterly*, 51: 590–621.

Klikauer, T. (2013) *Managerialism: A Critique of an Ideology*. Basingstoke: Palgrave.

Korczynski, M. (2003) 'Communities of Coping: Collective Emotional Labour in Service Work', *Organization*, 10(1): 55–79.

Korczynski, M. (2007) 'Service Work, Social Theory and Collectivism: A Reply to Brook', *Work, Employment and Society*, 21(3): 577–588.

Kronblad, C. (2020) 'How Digitization Changes our Understanding of Professional Service Firms', *Academy of Management Discoveries*, 6(3): 436-454.

Kuhn, G. (2001) 'Circadian Rhythm, Shift Work and Emergency Medicine', *Annals of Emergency Medicine*, 37(1): 88–98.

Kusenbach, M. (2003) 'Street Phenomenology: The Go-along as Ethnographic Research Tool', *Ethnography*, 4(3): 455–485.

Kyed, M. (2016) 'Masculinity, Emotions and "Communities of Relief" among Male Emergency Medical Technicians', in Ericson and Mellström, U. (eds), *Masculinities, Gender Equality and Crisis Management*, Abingdon: Routledge, pp. 34–46.

Kyed, M. (2020) 'Doing Care Work on the Fly – Exploring the Unnoticed Socio-emotional Skills of Male Ambulance Staff', *Sociology of Health and Illness*, 42(3): 433–448.

Lapsley, I. (2009) 'New Public Management: The Cruellest Invention of the Human Spirit?', *Abacus*, 45(1): 1–21.

Larson, M. S. (2013/1977) *The Rise of Professionalism: Monopolies of Competence and Sheltered Markets*. New Brunswick: Transaction Publishers.

Lawn, S., Roberts, L., Willis, E., Couzner, L., Mohammadi, L., and Goble, E. (2020) 'The Effects of Emergency Medical Service Work on the Psychological, Physical, and Social Well-being of Ambulance Personnel: A Systematic Review of Qualitative Research', *BMC Psychiatry*, 20(348) https://doi.org/10.1186/s12888-020-02752-4

Learmonth, M. (2019) 'Rethinking the New "Leadership" Mainstream: An Historical Perspective from the National Health Service', in Wankhede, P., McCann, L., and Murphy, P.

(eds), *Critical Perspectives on the Management and Organization of Emergency Services*. Abingdon: Routledge, pp. 235–253.

Learmonth, M. and Morrell, K. (2019) *Critical Perspectives on Leadership: The Language of Corporate Power*. Abingdon; Routledge.Lee, R. M. (1995) *Dangerous Fieldwork*. London: Sage.

Leicht, K. (2016) 'Market Fundamentalism, Cultural Fragmentation, Post-Modernism, Scepticism, and the Future of Professional Work', *Journal of Professions and* Organization, 3: 103–117.

Leicht, K. and Fennell, M. L. (1997) 'The Changing Organizational Context of Professional Work', *Annual Review of Sociology*, 23: 215–231.

Leicht, K. T. and Fennell, M. L. (2001) *Professional Work: A Sociological Approach*. Oxford: Blackwell.

Lennie, S. J., Crozier, E., and Sutton, A. (2020) 'Robocop: The Depersonalisation and Police Officers and Their Emotions: A Diary Study of Emotional Labour and Burnout in Front Line British Police Officers', *International Journal of Law, Crime and Justice*, 61: 100365.

Lewis, D. (2017) *Bullying and Harassment at South East Coast Ambulance Service*. Independent report commissioned by South East Coast Ambulance Service. Available at: http://www.secamb.nhs.uk/about_us/news/2017/bullying__harassment_report.aspx

Liaschenko, J. and Peter, E. (2004) 'Nursing Ethics and Conceptualizations of Nursing: Profession, Practice and Work', *Journal of Advanced Nursing*, 46(5): 488–495.

Liljegren, A. and Saks, M. (2017) 'Introducing Professions and Metaphors', in Liljegren, A. and Saks, M. (eds), *Professions and Metaphors: Understanding Professions in Society*. Abingdon: Routledge, pp. 1–9.

Lipsky, M. (2010 [1980]) *Street-Level Bureaucracy: Dilemmas of the Individual in Public Services*. New York: Russell Sage Foundation.

Loftus, B. (2009) *Police Culture in a Changing World*. Oxford: Oxford University Press.

Loftus, B. (2010) 'Police Occupational Culture: Classic Themes, Altered Times', *Policing and Society*, 20(1): 1–20.

Lois, J. (2003) *Heroic Efforts: The Emotional Culture of Search and Rescue Volunteers*. New York: New York University Press.

Lopez, S. H. (2010) 'Workers, Managers, and Customers: Triangles of Power in Work Communities', *Work and Occupations*, 37(3): 251–271.

Lowe, R. A. and Abbuhl, S. B. (2001) 'Appropriate Standards for 'Appropriateness' Research.' *Annals of Emergency Medicine*, 37(6): 629–632.

Lyng, S. (1990) 'Edgework: A Social Psychological Analysis of Voluntary Risk Taking', *American Journal of Sociology*, 95(4): 851–886.

Lyng, S. ed. (2005) *Edgework: The Sociology of Risk-Taking*. London: Routledge.

Lyng, S. (2014) 'Action and Edgework: Risk Taking and Reflexivity in Late Modernity', *European Journal of Social* Theory, 17(4): 443–460.

Mackintosh, N., Humphrey, C., and Sandall, J. (2014) 'The Habitus of 'Rescue' and Its Significance for Implementation of Rapid Response Systems in Acute Health Care', *Social Science & Medicine*, 120: 233–242.

Maguire, B. J., O'Meara, P. F., Brightwell, R. F., O'Neill, B. J., and Fitzgerald, G. J. (2014) 'Occupational Injury Risk among Australian Paramedics: An Analysis of National Data', *Medical Journal of Australia*, 200(8): 477–480.

Maitlis, S. and Christianson, M. (2014) 'Sensemaking in Organizations: Taking Stock and Moving Forward', *Academy of Management Annals*, 8(1): 57–125.

Mannon, J. M. (1992) *Emergency Encounters: EMTs and Their Work*. Boston, MA: Jones and Bartlett.

Manolchev, C. and Lewis, D. (2021) 'A Tale of Two Trusts: Case Study Analysis of Bullying and Negative Behaviours in the UK Ambulance Service', *Public Money & Management*, online early.

Marinetto, M. (2011) 'A Lipskian Analysis of Child Protection Failures from Victoria Climbie to Baby P: A Street-Level Re-Evaluation of Joined-Up Governance', *Public Administration*, 89(3): 1164–1881.

Mars, B., Hird, K., James, C., and Gunnell, D. (2020) 'Suicide among Ambulance Service Staff: A Review of Coroner and Employment Records', *British Paramedic Journal*, 4(4): 10–15.

Martin, B. (2016) 'What's happening to our universities', *Prometheus*, 34(1): 7–24.

Maunder, R. G., Halpern, J., Schwartz, B., and Gurevich, M. (2012) 'Symptoms and Responses to Critical Incidents in Paramedics Who Have Experienced Childhood Abuse and Neglect', *Emergency Medicine Journal*, 29: 222–227.

Maynard-Moody, S. and Musheno, M. (2003) *Cops, Teachers, Counselors: Stories from the Front Lines of Public Service*. Ann Arbor, MI: University of Michigan Press.

McCann, L. (2016) '"Management Is the Gate": But to Where? Rethinking Robert McNamara's "Career Lessons"', *Management and Organizational History*, 11(2): 166–188.

McCann, L. and Granter, E. (2019) 'Beyond 'Blue-Collar Professionalism': Continuity and Change in the Professionalization of Uniformed Emergency Services Work', *Journal of Professions and Organization*, 6(2): 213–232.

McCann, L., Granter, E., Aroles, J., and Hyde, P. (2020) '"Upon the Gears and Upon the Wheels": Terror Convergence and Total Administration in the Neoliberal University', *Management Learning*, 51(4): 431–451.

McCann, L., Granter, E., Hassard, J., and Hyde, P. (2015) '"You Can't Do Both – Something Will Give": Limitations of the Targets Culture in Managing UK Health Care Workforces', *Human Resource Management*, 54(5): 773–791.

McCann, L., Granter, E., Hyde, P., and Hassard, J. (2013) 'Still Blue-Collar After All These Years? An Ethnography of the Professionalization of Emergency Ambulance Work', *Journal of Management Studies*, 50(5): 750–776.

McCormack, B., McCance, T., Bulley, C., Brown, B., McMillan, A., and Martin, S. eds (2021) *Fundamentals of Person-Centred Healthcare Practice: A Guide for Healthcare Students*. Hoboken, NJ: Wiley-Blackwell.

Metz, D.L. (1981) *Running Hot: Structure and Stress in Ambulance Work*. Cambridge, MA: Abt Books.

Mildenhall, J. (2012) 'Occupational Stress, Paramedic Informal Coping Strategies: A Review of the Literature', *Journal of Paramedic Practice*, 4(6): 318–328.

Mildenhall, J. (2019) 'Protecting the Mental Health of UK Paramedics', *Journal of Paramedic Practice*, 11(1): 6–7.

Mildenhall, J. (2021) 'Paramedics' Lived Experiences of Post-Incident Traumatic Distress and Psychosocial Support: An Interpretative Phenomenological Study', in Murray, E., and Brown, J. (eds), *The Mental Health and Wellbeing of Healthcare Practitioners: Research and Practice*. New York: Wiley, pp. 54–71.

Mind (2018) *Blue Light Programme Research Summary 2016–18*. London: Mind. Available at: https://www.mind.org.uk/media-a/4597/blue-light-programme-research-summary_2016-to-18_online.pdf (accessed 25 June 2021).

Monrouxe, L. V., Rees, C. E., and Hu, W. (2011) 'Differences in Medical Students' Explicit Discourses of Professionalism: Acting, Representing, Becoming', *Medical Education*, 45(6): 585–602.

Moskos, P. (2009) *Cop in the Hood: My Year Policing Baltimore's Eastern District*. Princeton: Princeton University Press.

Mueller, T. (2020) *Crisis of Conscience: Whistleblowing in an Age of Fraud*. London: Atlantic.

Murphy, R. (1988) *Social Closure: The Theory of Monopolization and Exclusion*. Oxford: Clarendon Press.

Murray, E., (2019) 'Moral Injury and Paramedic Practice', *Journal of Paramedic Practice*, 11(10): 424–425.

Muzio, D., Aulakh, S., and Kirkpatrick, I. (2020) *Professional Occupations and Organizations*. Cambridge: Cambridge University Press.

National Audit Office (2017) *NHS Ambulance Services*.London: NAO.

Nelson, B. J. (1997) 'Work as a Moral Act: How Emergency Medical Technicians Understand Their Work', in Barley, S. R. and Orr, J. E. (eds), *Between Craft and Science: Technical Work in U.S. Settings*. Ithaca, New York: Cornell University Press, pp. 154–184.

Nelson, B. J. and Barley, S. R. (1997) 'For Love or Money? Commodification and the Construction of an Occupational Mandate', *Administrative Science Quarterly*, 42(4): 619–653.

Nelson, P. A., Cordingley, L., Kapur, N., Chew-Graham, C. A., Shaw, J., Smith, S., McGale, B., and McDonnell, S. (2020) '"We're the First Port of Call" – Perspectives of Ambulance Staff on Responding to Death by Suicide: A Qualitative Study', *Frontiers in Psychology*, 11: 722

Newton, A. (2019) 'Quo Vadis: Eight Possible Scenarios for Changes in the Ambulance Services', in Wankhade, P., McCann, L., and Murphy, P. (eds), *Critical Perspectives on the Management and Organization of Emergency Services*. Abingdon: Routledge, pp. 70–90.

Newton, A., Hunt, B., and Williams, J. (2020) 'The Paramedic Profession: Disruptive Innovation and Barriers to Further Progress', *Journal of Paramedic Practice*, 12(4): 138–148.

Nichols, T. (2017) *The Death of Expertise: The Campaign against Established Knowledge and Why It Matters*. Oxford: Oxford University Press.

Nicklich, M., Braun, T., and Fortwengel, J. (2020) 'Forever a profession in the making? The intermediate status of project managers in Gemany', *Journal of Professions and Organization*, 7(3): 374–394.

Noordegraaf, M. (2007) 'From "Pure" to "Hybrid" Professionalism: Present-day Professionalism in Amigious Public Domains', *Administration & Society*, 39(6): 761–785.

Noordegraaf, M. (2015) 'Hybrid Professionalism and Beyond. (New) Forms of Public Professionalism in Changing Organizational and Societal Contexts', *Journal of Professions and Organizations*, 2(2): 187–206.

Noordegraaf, M. (2016) 'Reconfiguring Professional Work: Changing Forms of Professionalism in Public Services', *Administration and Society*, 48(7): 783–810.

Noordegraaf, M. (2020) 'Protective or Connective Professionalism? How Connected Professionals Can (Still) Act as Autonomous and Authoritative Experts', *Journal of Professions and Organizations*, 7(2): 205–223.

Nurok, M. and Henckes, N. (2009) 'Between Professional Values and the Social Valuation of Patients: The Fluctuating Economy of Pre-hospital Emergency Work', *Social Science & Medicine*, 68(3): 504–510.

O'Connor, S. D. (1998) 'Professionalism', *Washington University Law Quarterly*, 76(5): 5–13.

O'Neill, M. (2001) 'Participation or Observation? Some Practical and Ethical Dilemmas', in Gellner, D. N and Hirsh, E. (eds), *Inside Organizations: Anthropologists at Work*. Oxford: Berg, pp. 223–230.

O'Toole, M. and Calvard, T. (2019) '"I've Got Your Back": Danger, Volunteering and Solidarity in Lifeboat Crews', *Work, Employment & Society*, 34(1): 73–90.

Ordóñez, L.D., Schweitzer, M.E., Galinsky, A.D. and Bazerman, M.H. (2009) 'Goals Gone Wild: The Systematic Side Effects of Overprescribing Goal Setting', *Academy of Management Perspectives*, 23(1): 6–16.

Orr, J. E. (1996) *Talking about Machines: An Ethnography of a Modern Job*. Ithaca: Cornell University Press.

Orr, J. E. (2006) 'Ten Years of Talking About Machines', *Organization Studies*, 27(10): 1805–1820.

Oshige, K., Kawatami, C., Kubota, K., and Tochikubo, M. D. (2008) 'A Contingent Valuation Study of the Appropriate User Price for Ambulance Service', *Academic Emergency Medicine*, 12(10): 932–940.

Palazzo, F. F., Warner, O. J., Harron, M., and Sadana, A. (1998) 'Misuse of the London Ambulance Service: How Much and Why?', *Emergency Medicine Journal*, 15: 368–370.

Palmer, C. E. (1983) '"Trauma Junkies" and Street Work: Occupational Behavior of Paramedics and Emergency Medical Technicians', *Urban Life*, 12(2): 162–183.

Palmer, L. (2012) '*Cranked* Masculinity: Hypermediation in Digital Action Cinema', *Cinema Journal*, 51(4): 1–25.

Perkin, H. (2002) *The Rise of Professional Society: England since 1880*. London: Routledge.

Petrie, K., Milligan-Saville, J., Gayed, A., Deady, M., Phelps, A., Dell, L., Forbes, D., Bryant, R.A., Calvo, R.A., Glozier, N., and Harvey, S.B. (2018) 'Prevalence of PTSD and common mental disorders amongst ambulance personnel: a systematic review and meta-analysis', *Social Psychiatry and Psychiatrice Epidemiology*, 53: 897–909.

Pilbery, R. and Lethbridge, K. (2015) *Ambulance Care Essentials*. Bridgwater: Class Professional Publishing.

Player, D. and Barbour-Might, D. (1988) *Health for All: The Practical Socialism of the National Health Service*. London: Tribune.

Pogrebin, M. R. and Poole, E. D. (1988) 'Humor in the Briefing Room: A Study of the Strategic Uses of Humor among Police', *Journal of Contemporary Ethnography*, 17(2): 183–210.

Ponnert, L. and Svensson, K. (2016) 'Standardisation – the End of Professional Discretion?', *European Journal of Social Policy*, 19(3–4): 586–599.

Powell, A. E. and Davies, H. T. O. (2012) 'The Struggle to Improve Patient Care in the Face of Professional Boundaries', *Social Science & Medicine*, 75: 807–814.

Power, M. (1997) *The Audit Society: Rituals of Verification*. Cambridge: Cambridge University Press.

Pringle, L. (2009) *Blue Lights and Long Nights*. London: Corgi.

Rafi Khan, S. (2018) 'Reinventing Capitalism to Address Automation: Sharing Work to Secure Employment and Income', *Competition & Change*, 22(4): 343–362.

Ramthun, A. J. and Matkin, G. S. (2014) 'Leading Dangerously: A Case Study of Military Teams and Shared Leadership in Dangerous Environments', *Journal of Leadership and Organizational Studies*, 21(3): 244–256.

Reed, K. (2020) 'Miscarriage, SUDI and neonatal death: paramedic experience and practice', *Journal of Paramedic Practice*, 12(12): 472-477.

Reed, M. I. (2018) 'Elites, Professions, and the Neoliberal State: Critical Points of Intersection and Contention', *Journal of Professions and Organization*, 5: 297–312.

Reiner, R. (2010) *The Politics of the Police*, Oxford: Oxford University Press.

Reynolds, T. (2009) *Blood, Sweat and Tea: Real-Life Adventures in an Inner-City Ambulance*. London: Friday Project

Reynolds, T. (2010) *More Blood, More Sweat, and Another Cup of Tea*. London: Friday Project.

Riccucci, N. M. (2005) *How Management Matters: Street-level Bureaucrats and Welfare Reform*. Washington, D.C.: Georgetown University Press.

Saks, M. (2020) 'Introduction: Support Workers and the Health Professions', in Saks, M. (ed.), *Support Workers and the Health Professions in International Perspective: The Invisible Providers of Health Care*. Bristol: Policy Press, pp.1–14.

Saks, M. (2021) *Professions: A Key Idea for Business and Society*. Abingdon: Routledge.

Salaman, G. (1971) 'Two Occupational Communities: Examples of a Remarkable Convergence of Work and Non-Work', *The Sociological Review*, 19(3): 389–407.

Satyamurti, C. (1981) *Occupational Survival*. Oxford: Basic Blackwell.

Scott, C. and Tracy, S. J. (2007) 'Riding Fire Trucks & Ambulances with America's Heroes', in Drew, S. K., Mills, M. B. and Gassaway, B. M. (eds), *Dirty Work: The Social Construction of Taint*. Waco, TX: Baylor University Press, pp. 55–76.

Scott, W. R. (1982) 'Managing Professional Work: Three Models of Control for Health Organizations', *Health Services Research*, 17: 213–240.

Scott, W. R. (2008) 'Lords of the Dance: Professionals as Institutional Agents', *Organization Studies*, 29(2): 219–238.

Seim, J. (2017) 'The Ambulance: Towards a Labor Theory of Poverty Governance', *American Sociological Review*, 82(3): 451–475.

Seim, J. (2020) Bandage, Sort, and Hustle: Ambulance Crews on the Front Lines of Human Suffering. Oakland, CA: University of California Press.

Senate Education and Employment Committees (2019) *The People Behind 000: Mental Health of Our First Responders*. Canberra: Parliament of Australia.

Shay, J. (1994) *Achilles in Vietnam: Combat Trauma and the Undoing of Character*. New York: Scribner.

Shay, J. (2014) 'Moral Injury', *Psychoanalytic Psychology*, 31(2): 182–191.

Simon, D. (2006) *Homicide*. Edinburgh: Canongate.

Smith, A. (2002) 'Leaning about Reflection', *Journal of Advanced Nursing*, 28(4): 891–898.

Snooks, H., Wrigley, H., George, S., Thomas, E., Smith, H., and Glasper, A. (1998) 'Appropriateness of the Use of Emergency Ambulances', *Journal of Accident and Emergency Medicine*, 15: 212–218.

Soeters, J. (2018) *Sociology and Military Studies: Classical and Current Foundations*. Abingdon: Routledge.

Sommer Harrits, G. (2019) 'Street-Level Bureaucracy Research and Professionalism', in Hupe, P., ed., (2019) *Research Handbook on Street-Level Bureaucracy*. Aldershot: Edward Elgar, pp. 193–208

Stein, M. (2021) 'The Lost Good Self: Why the Whistleblower Is Hated and Stigmatized'. *Organization Studies*, 42(7): 1167–1186.

Stewart, R. (2018) *Evidence-based Management: A Practical Guide for Health Professionals*. London: CRC Press.

Strathern, M. (2000) 'Introduction: New Accountabilities', in Strathern, M. (ed.), *Audit Cultures: Anthropological Studies in Accountability, Ethics and the Academy*. London: Routledge, pp. 1–18.

Styhre, A. (2017) *Precarious Professional Work: Entrepreneurialism, Risk and Economic Compensation in the Knowledge Economy*. Cham, Switzerland: Palgrave.

Susskind, R. and Susskind, D. (2015) *The Future of the Professions: How Technology Will Transform the Work of Human Experts*. Oxford: Oxford University Press.

Tangherlini, T. (1998) *Talking Trauma: A Candid Look at Paramedics through Their Tradition of Tale-Telling*. Jackson, MS: University Press of Mississippi.

Tangherlini, T. (2000) 'Heroes and Lies: Storytelling among Paramedics', *Folklore*, 111:43–66.

Taylor, R. (2013) *God Bless the NHS: The Truth Behind the Current Crisis*. London: Faber & Faber.

ten Hoeve, Y. and Jansen, G. (2013) 'The Nursing Profession: Public Image, Self-Concept and Professional Identity. A Discussion Paper', *Journal of Advanced Nursing*, 70(2): 295–309.

The Police Foundation (2013) *Roadcraft: The Police Driver's Handbook*. London: The Stationery Office.

Thomas, P., McArdle, L., and Saundry, R. (2020) 'Introduction to the Special Issue: The Enactment of Neoliberalism in the Workplace: The Degradation of the Employment Relationship'. *Competition & Change*, 24(2): 105–113.

Tolich, M. (2010) 'A Critique of Current Practice: Ten Foundational Guidelines for Autoethnographers', *Qualitative Health Research*, 20(12): 1599–1610.

Tomkins, L. and Bristow, A. (2021) 'Evidence-Based Practice and the Ethics of Care: 'What Works' or 'What Matters'?', *Human Relations*, online early

Turnbull, P. and Wass, V. (2015) 'Normalizing extreme work in the Police Service? Austerity and the inspecting ranks', *Organization*, 22(4): 512–529.

van der Gaag, A., Jago, R., Austin, Z., Zasada, M., Banks, S., Gallagher, A., and Lucas, G. (2018) 'Why Do Paramedics Have a High Rate of Self-Referral?' *Journal of Paramedic Practice*, 10(5): 205–210.

van Hulst, M. and Tsoukas, H. (2021) 'Understanding Extended Narrative Sensemaking: How Police Officers Accomplish Story Work', *Organization*, online early.

van Maanen, J. and Barley, S. R. (1984) 'Occupational Communities: Culture and Control in Organizations. *Research in Organizational* Behavior, 6: 287–365.

Vinzant, J. C. and Cruthers, L. (1998) *Street-Level Leadership: Discretion and Legitimacy in Front-Line Public Service*. Washington, DC: Georgetown University Press.

Waddington, P. A. J. (1999) 'Police (Canteen) Sub-culture. An Appreciation', *British Journal of Criminology*, 39(2): 287–309.

Walder, L. (2020) *Fighting for Your Life: A Paramedic's Story*. London: John Blake.

Wankhade, P. (2011) 'Performance Measurement and the UK Emergency Ambulance Service: Unintended Consequences of the Ambulance Response Time Targets', *International Journal of Public Sector Management*, 24(5): 384–402.

Wankhade, P. (2012) 'Different Cultures of Management and Their Relationships with Organizational Performance: Evidence from the UK Ambulance Service', *Public Money & Management*, 32(5): 381–388.

Wankhade, P., McCann, L., and Murphy, P. (2019) 'Introduction' in Wankhade, P., McCann, L., and Murphy, P. (2019) *Critical Perspectives on the Management and Organization of Emergency Services*. Abingdon: Routledge, pp. 1–11.

Warden, T. (2013) 'Feet of Clay: Confronting Emotional Challenges in Ethnographic Experience', *Journal of Organizational Ethnography*, 2(2): 150–172.

Wastell, D., White, S., Broadhurst, K., Peckover, S., and Pithouse, A. (2010) 'Children's Services in the Iron Cage of Performance Management: Street-Level Bureaucracy and the Spectre of Svejkism'. *International Journal of Social Welfare*, 19(3): 310–320.

Watson, T. (2011) 'Ethnography, Reality and Truth: The Vital Need for Studies of "How Things Work" in Organisations and Management', *Journal of Management Studies*, 48(1): 202–217.

Wharton, K. (2017) *Emergency Admissions: Memoirs of an Ambulance Driver*. London: 4th Estate.

White, K. M. (2014) 'Professionalism and Nursing – A Quest or an Accomplishment?', in DeAngelis, C. D. (ed.), *Patient Care and Professionalism*. Oxford: Oxford University Press, pp. 77–98.

Wilensky, H. L. (1964) 'The Professionalization of Everyone?', *American Journal of Sociology*, 70(2): 137–158.

Willis, S. (2015) 'Communication Skills for the Pre-hospital Professional' in Willis, S. and Dalrymple, R. (eds), *Fundamentals of Paramedic Practice: A Systems Approach*. Oxford: Wiley Blackwell, pp. 29–44

Willis, S. and Dalrymple, R. eds (2015) *Fundamentals of Paramedic Practice: A Systems Approach*. Oxford: Wiley Blackwell.

Zacka, B. (2017) *When the State Meets the Street: Public Service and Moral Agency*. Boston, MA: The Belknap Press of Harvard University Press.

Zussman, R. (2016) 'Alice's Adventures in Wonderland: *On the Run* and Its Critics'. *Society*, 53: 436–443.

Index